T0229858

List of Figures

List of Tables

Foreword

I am delighted to introduce this interesting book on large-scale simulation technologies. By simulation, I mean a computer-based simulation, which is a computer application that attempts to formulate the dynamics of a real or imagined system in a certain abstract level. With the fast increasing complexity of problem domains, people often need to identify a particularly large and complex system exactly as something like "a modern battlefield," "a global supply chain," and "a global climate model," etc.

For more than a decade, research in this direction has been very productive with a large number of publications of high quality. This direction has been put to more and more successful applications in many diverse domains. Undoubtedly, a book can play a significant role in maturing a research direction. I appreciate the three authors' efforts to unify the field by bringing in disparate topics via a series of solid research work of their own and other exciting papers.

The authors' interpretations of the architectures for large-scale simulation, simulation cloning, fault tolerance, synchronization mechanisms, and agent-based models are of great value for undergraduate/postgraduate students, researchers, and engineers who are exploring the simulation discipline or the subdomain of modeling and simulation of complex systems. The book provides a comprehensive background knowledge foundation that identifies very helpful ideas, contemporary methods, and solid examples of the research work. For researchers and instructors of the modeling and simulation of complex systems, this book reviews much they have been familiar with but certainly gives them new insights into issues and methods, and assures their recognition of the latest advances in the findings.

I was inspired by this book project at the very beginning. The book now appears to be a much better manuscript. The resulting work is really thoughtful, creative, and comprehensive. I highly recommend this book.

Lawrence T. Yang
Department of Computer Science
St. Francis Xavier University

Preface

A computer-based simulation model formulates a real-life or hypothetical system that is executable over computers so that the simulated system can be analyzed to see how it evolves over time. Computer simulation has become an essential part of modeling many natural systems in physics, chemistry and biology, and human systems in economics and social science (the computational sociology) as well as in engineering to gain insight into the operation of those systems.

Large-scale simulation empowered by high-performance computing enables a complex simulation program to be executed on a parallel or distributed computer system. A large-scale distributed simulation may be constructed by linking together existing simulation models across multiple locations to achieve reusability and interoperability. The past decade has witnessed an explosion of interest in large-scale distributed simulations due to the increasing scale, resolution, and complexity of systems to be studied. The culmination of these activities includes the advent of the High Level Architecture (HLA), an IEEE standard to facilitate interoperability and reuse among simulation models. Recently, with the advance of computing technologies, like Grids, modern cyberinfrastructure, Web services, and multi-core/many-core architectures, more research efforts have been performed on large-scale simulation techniques. However, there is no specific book that reflects the latest research advances in this field. This book is devoted to filling this gap and to summarizing the authors' research achievements in connection with important research issues for large-scale simulations since 2000. Most of the research results in this book have been published in top-tier, peer-reviewed, scientific conference proceedings and journals. By complementing an introduction and foundation of the research, this book organizes the study in a clear manner: model $\rightarrow$ algorithm $\rightarrow$ applications. This book discusses the advanced models and algorithms for large-scale distributed simulation. It is organized in four parts:

- **Part I:** Fundamentals of the study of large-scale simulation, including background, HLA/RTI, and related work

- **Part II:** Middleware and software architecture for large-scale simulations, for example, decoupled federate architecture, fault-tolerant mechanism, Grid-enabled simulation, and federation community

- **Part III:** Mechanisms that support quick evaluation of alternative scenarios, which mainly consist of simulation cloning methods and algorithms

- **Part IV:** Important applications of large-scale simulation for the study of social phenomena using distributed computing technologies and many-core architecture

The authors would like to express their deep gratitude to Professor Stephen J. Turner. This erudite and perspicacious scholar gave us massive support in faith, with great patience, throughout the course of most research projects in connection with this book. Special thanks to Professor Wentong Cai of Nanyang Technological University, for his unreserved contributions and ideas. Through him, we have acquired countless knowledge in this area.

Dan Chen, Lizhe Wang, and Jingying Chen

About the Authors

Dr. Dan Chen is a full professor and the director of the Scientific Computing Lab with the School of Computer Science, China University of Geosciences, China. His research interests include computer-based modeling and simulation, high-performance computing, and neuroinformatics. Dr. Chen has published more than 60 research papers in these research areas. He was a Joint Higher Education Funding Council for England (HEFCE) Research Fellow at the University of Birmingham (UoB), and the University of Warwick, U.K. Before he joined UoB, he was an Associate Research Fellow at the Singapore Institute of Manufacturing Technology, Singapore. He received a B.Sc. degree from Wuhan University, China, an M.Eng. degree from the Huazhong University of Science and Technology, China, and another M.Eng. degree and his Ph.D. degree from Nanyang Technological University, Singapore.

Dr. Lizhe Wang is a full professor at the Center for Earth Observation and Digital Earth (CEODE), the Chinese Academy of Sciences. Before he joined the CEODE, Dr. Wang was a research scientist and principal research engineer at the Pervasive Technology Institute, Indiana University. Dr. Wang was awarded the Research Innovation Award in 2010 by the HP Lab. Dr. Wang has published more than 70 papers and 6 books. Dr. Wang has served as program chair/general chair for more than 10 conferences/workshops, and program committee member for more than 50 conferences/workshops. Dr. Wang's research interests include high-performance computing, Grid computing and Cloud computing. Currently Dr. Wang is leading a group at the Division of Data Technology and carries out his research on high-performance storage service and massive data processing for remote sensing.

Dr. Jingying Chen is a full professor with the National Engineering Center for E-Learning, Central China Normal University, China. Her research interests include image processing, computer vision, pattern recognition, and human–machine interface. She was a post-doctor in INRIA, France, and a Research Fellow with the University of St. Andrews and the University of Edinburgh, U.K. She received her Bachelor's and Master's degrees from the Huazhong University of Science and Technology, China, in 1996 and 1998, respectively. She received her Ph.D. degree from the School of Computer Engineering, Nanyang Technological University, Singapore, in 2002.

Acknowledgments

This work was supported in part by National Natural Science Foundation of China (grant No. 60804036), the Specialized Research Fund for the Doctoral Program of Higher Education (grant No. 20110145110010), the Programme of High-Resolution Earth Observing System (China), the Fundamental Research Funds for the Central Universities (CUGL100608, CUG, Wuhan), the Program for New Century Excellent Talents in University (grant No. NCET-11-0722), Natural Science Foundation of Hubei Province of China (grant No. 2011CDB159), and the CNU Talent Programme (grant No. 120005030223).

Dan Chen gratefully acknowledges support from the Birmingham-Warwick Science City Research Alliance. Lizhe Wang's work in this book was funded by "One Hundred Talents" Programme of Chinese Academy of Sciences.

Part I

Fundamentals

1

Introduction

CONTENTS

1.1   Background

Simulation technology is commonly used to study physical systems or imagined systems [41]. A complex system can be modeled as a simulation program (or programs), which simulates the system's behavior on computers. The simulations can be used to perform what-if analysis on the simulated system, to predict the different results of varying key parameters/estimates, or to examine a series of solutions prior to making the final implementation decision.

Sometimes it is unmanageable, expensive, and/or risky to construct a real system for analysis. For example, people want to explore the eruption process of an active volcano. Astronomers attempt to discover the evolution of the Universe. Researchers may have to choose to use simulation to emulate the activities of these systems. Simulation technology is safer, more feasible and flexible, and possibly much less costly in practice. Simulation can overcome the inconvenience incurred by the span in time and geographic distribution. Days or even years of activity can be executed in a very short time in simulation executions. Thus, simulation can provide prediction to support decision-making. Furthermore, the external environment is unlikely to impact the execution of simulations while it often limits the operation of real systems.

Nowadays, simulation technology has a wide application spectrum, from training to gaming, from scientific research to business operation, and from military to civilian purposes. It is involved in a variety of human and natural activities. Battlefield simulation environments over networked computers have already replaced the sand table for plotting military operations. Flight simulators can help in training pilots safely and efficiently from the Boeing series to Longbow Apache. Networked multimedia together with simulation facilitates the creation of new exciting automated learning environments [94]. Many PC games like Command&Conquer construct an amazing virtual world for many people [68]. As for scientific research, algorithms and prototypes are often developed based on simulation models. In practice, system designers often exploit simulations to verify the proposed designs; thus they can perform their work conveniently and efficiently [32]. Industry and business collaborations also benefit from simulation for decision support and productivity enhancement [79]. Simulation can help companies facilitate the design, evaluation, and optimization of their daily operations.

There are different types of simulations, such as continuous simulation and discrete event simulation, parallel and distributed simulation. In continuous simulation, states change continuously over simulation time, but in discrete event simulation, system states change at discrete points in time [41]. The events represent the system activities or changes of system state. An associated timestamp denotes the simulation time at which an event occurs. For example, in a supply-chain simulation [44], the arrival of a customer order can be modeled as an event.

The simulation of a complex system can be divided into smaller components to form a parallel simulation. Each component, known as a Logical Process (LP), models a subset of the system. Parallel simulation concurrently executes LPs over multiple processors to reduce execution time.

Distributed simulation is an important technology that enables a simulation program to be executed on a distributed computer system. A large-scale simulation may be constructed by linking together existing simulation models at possibly different locations. For example, distributed simulation technology meets the pressing need of supply-chain simulation, as a supply-chain often involves multiple companies across enterprise boundaries [44]. Each of the companies participating in a supply-chain simulation may already have their existing simulation models, which can be developed independently based on heterogeneous platforms [108]. Distributed simulation technology based on a common standard can facilitate the interoperability among these distributed simulation models.

Distributed simulation has been the focus of the U.S. defense industry for over a decade. These defense-related efforts originated with SIMNET [93] and evolved into the Distributed Interactive Simulation (DIS) protocol initiative [28, 64]. To meet the demand for standardization on a higher abstraction level, the Defense Modeling and Simulation Office (DMSO) sponsored the definition and development of the High Level Architecture (HLA) for modeling and simulation (M&S). The HLA defines an architecture for reuse and interoperation of simulations. The HLA supports component-based simulation development; the components are referred to as simulation federates [72]. Thus, a set of simulation federates, possibly developed independently, can be linked together to form a large federation. The HLA has been approved as an open standard through the Institute of Electrical and Electronic Engineers (IEEE) [66]. The Runtime Infrastructure (RTI) is the software to support an HLA-compliant distributed simulation.

In recent years, we have witnessed the rapidly increasing Complexity and scales of problem domains; there exists a pressing need for technologies to sustain simulations of large sizes. With the advance of computing technologies, like Grids, modern cyberinfrastructure, Web services, and multi-core/many-core architectures, more research efforts have been performed on large-scale simulation techniques.

1.2   Organization of the Book

Including the introduction chapter, this book contains twelves chapters in total. Chapter 2 introduces the basics of HLA and RTI services. This chapter also introduces the work related to large-scale simulations, including simulation cloning, fault tolerance, and synchronization in federation communities.

Chapter 3 introduces a novel Decoupled federate architecture to enable federate cloning at runtime. The decoupled architecture ensures the correct replication of federates and facilitates fault tolerance at the RTI level. This chapter gives preliminary benchmark experiments to study the extra overhead incurred by the Decoupled federate architecture against the normal federate. This chapter also discusses the exploitation of the Decoupled federate architecture used in distributed simulation cloning. A fault-tolerant model is proposed to provide runtime robustness. A Web/Grid-enabled architecture is introduced to support the use of HLA in a Web/Grid environment. The advantages of the decoupled architecture in providing load balancing are also discussed.

Chapter 4 presents a generic model and software framework that supports fault-tolerant large simulations based on the HLA. The framework deals with failure with a dynamic substitution approach. A sender-based method is designed to ensure reliable in-transit message delivery, which is coupled with a fossil collection algorithm. Experiments have been carried out to validate and benchmark the fault-tolerant federates.

Chapter 5 details a synchronization mechanism for federation community networks. A mathematical proof is given to examine the mechanism's correctness. The synchronization mechanism is suited for various types of federation community networks and supports the reusability of legacy federates. It also allows simulation users to benefit from both the Grid computing technologies and federation community approach.

Chapter 6 introduces the foundation theory of distributed simulation cloning. Basic concepts of the technology are defined and critical research issues are identified. This chapter outlines various types of federate cloning and scenario representation in distributed simulations.

Chapter 7 discusses the issues involved in cloning distributed simulations based on the HLA, as well as proposing tentative solutions. Alternative solutions are compared from both the qualitative and quantitative points of view. A middleware approach is suggested to hide the implementation Complexity and provide reusability and user transparency to existing simulation federates. To measure the trade-off between Complexity and efficiency, this chapter introduces a series of experiments to benchmark various solutions at the RTI level.

Chapter 8 describes the method of using Data Distribution Management (DDM) to partition concurrent scenarios in a cloning-enabled distributed simulation. Two candidate solutions are introduced for managing scenarios and identifying each clone and scenario. This chapter details the design of these two solutions and analyzes their advantages and drawbacks. This chapter also discusses how the partitioning mechanism is implemented using a middleware method to address reusability issues.

Chapter 9 first describes the design and functionalities of the infrastructure enabling distributed. Second, this chapter details the algorithms of simulation cloning at Decision points, including state saving and replication, the method of coordinating and synchronizing clones, etc. Third, an incremental cloning mechanism is covered in detail, which is designed to replicate only those federates whose states will be affected. This mechanism employs an event checking algorithm for sharing federates in multiple scenarios, and supports correct HLA semantics. This chapter also summarizes the entire cloning mechanism, which makes clones of all federates at the same time.

The performance of cloning mechanisms is examined in Chapter 10, and their correctness is established as well. Experiments have been carried out to compare the execution time of entire cloning and incremental cloning-enabled federates using an example of a simple supply-chain simulation. Experimental results indicate that the proposed cloning technology provides correct distributed simulation cloning and can significantly reduce the execution time for evaluating different scenarios of distributed simulations. Moreover, the incremental cloning mechanism significantly surpasses entire cloning in terms of execution efficiency.

Chapter 11 introduces a simulation of evacuating thousands of pedestrians in a large urban area on top of a hierarchical Grid infrastructure. The Grid simulation infrastructure can facilitate a large crowd simulation comprising models of different grains and various types in nature.

In Chapter 12, we present a method based on the concept of vector field to formulate the way in which external stimuli may affect the behaviors of individuals in a crowd. This study adopts GPGPU to sustain massively parallel M&S of a confrontation operation involving a large crowd. Experimental results indicate that the approach is efficient in terms of both performance and energy consumption.

2

Background and Fundamentals

CONTENTS

This chapter first introduces the basic concepts of HLA and RTI. The HLA provides the underlying infrastructure for the mechanisms developed in this study. The second section describes existing replication techniques used in software engineering and distributed systems. The third section presents previous research work on cloning in simulation. Some specific features and requirements for the proposed distributed simulation cloning mechanisms are also presented.

2.1   High Level Architecture and Runtime Infrastructure

The Defense Modeling and Simulation Office (DMSO) sponsored and developed the HLA standard [81]. The DMSO provides a full-time focal point for information concerning Department of Defense modeling and simulation (M&S) activities. The HLA defines a software architecture for modeling and

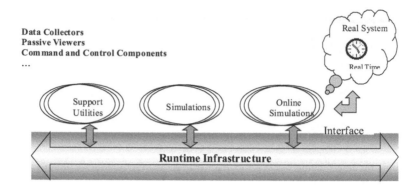

FIGURE 2.1
Functional view of an HLA federation.

simulation. The HLA is designed to provide reuse and interoperability of simulation components. The simulation components are referred to as federates. A simulation federation can be created to achieve some specific objective by combining simulation federates. The HLA supports component-based simulation development in this way [33]. A functional view of an HLA federation is illustrated in Figure 2.1.

The HLA federation is a collection of federates (observers, live participants, simulations, etc.) interacting with each other for a common purpose, for example war-gaming. These federates interact with each other with the support of the Runtime Infrastructure (RTI) and the use of a common **Federation Object Model** (FOM).

In the formal definition, the HLA standard comprises four main components: **HLA Rules, Object Model Template (OMT), Interface Specification** [66], and **Federation Development and Execution Process (FEDEP)** [67]. The HLA rules define the principles of HLA in terms of responsibilities that federates and federations must uphold. Each federation has an FOM, which is a common object model for the data exchanged between federates in a federation. The OMT defines the meta-model for all FOMs [72]. The HLA Interface Specification identifies the RTI services available to each federate and the functions each federate must provide to the federation. The FEDEP mainly defines a process framework for federation developers, including the necessary activities to build HLA federations.

As defined in the OMT, an "**object**" represents an entity with a distinct identity created in a federate, which belongs to a certain "**class**" and defines a set of named data called "**attributes**." An "**interaction**" is a collection of data sent at one time through the RTI to other federates, with accompanying "**parameters**," representing some occurrence in a federate. Objects and interactions are identified by unique handles assigned by the RTI. In the context of an HLA-compliant simulation, events can be updating attributes of an object or sending an interaction. Any federation model can be written in terms of objects and interactions [72], in which (1) any occurrence can be modeled as an interaction, (2) if an entity is to be modeled such that it has persistent state, the entity should be represented as an object.

The HLA is an architecture defining the rules and interface, whereas the RTI is the software conforming to the HLA standard, and is used to support a federation execution. Figure 2.2 gives an overview of an HLA federation and the RTI [72]. The RTI provides a set of services to the federates for data interchange and synchronization in a coordinated fashion. The RTI services are provided to each federate through its Local RTI Component (LRC) [1]. The RTI can be viewed as a distributed operating system providing services to support interoperable simulations executing in distributed computing environments [56]. The HLA exhibits three features [72]: a layered architecture, data abstraction architecture, and event-based architecture. The HLA architecture separates the simulation model from the infrastructure functions. Thus, all behavior specific to a given simulation model is in the federate, and the infrastructure contains functions generic to simulation interoperability. In a distributed federation, the LRC contains the network functions needed to facilitate distribution. The data abstraction architecture frees federates from retaining references to other federates. The event-based architecture provides implicit invocations to call back all receiving federates for an event and does not require the sending federate to know the receivers.

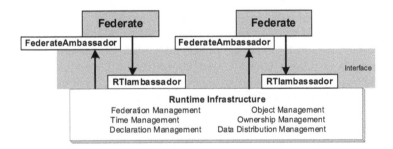

FIGURE 2.2
Overview of HLA federation and RTI.

A total of six service categories are defined in the specification, namely Federation Management, Declaration Management, Object Management, Ownership Management, Data distribution Management, and Time Management [1]. The definition and functionalities of these RTI services are given as follows.

Federation management services provide methods such as creating federations, joining federates to federations, observing federation-wide synchronization points, effecting federation-wide saves and restores, resigning federates from federations, and destroying federations.

Declaration management services include publication, subscription, and supporting control functions. Using services defined in this category, federates declare object class attributes or interactions (predefined in the OMT) that they intend to publish (produce) or subscribe (consume).

Object management services include instance registration and instance updates for the object producers. On the other hand, they provide instance discovery and reflection for the object consumers. They also support methods for sending and receiving interactions, controlling instance updates based on consumer demand, and other miscellaneous functions.

Ownership management services can be used to transfer ownership of instance attributes among federates. The ability to transfer ownership is intended to support the cooperative modeling of a given object instance across a federation. The services provided support both push and pull mechanisms for ownership transfer.

Data distribution management (DDM) services are employed by the data producers and consumers to assert properties of their data or to specify their data requirements respectively based on specified regions. The RTI then distributes the data from the producers to the consumers based on the matches between the properties and the requirements. This management controls the efficient routing of class attributes and interactions via the RTI.

Time management is concerned with the mechanisms for controlling the advancement of each federate along the federation time axis. The services in this group synchronize event delivery among federates. Time advances are coordinated by the RTI with object management services so that information is delivered to federates in a causally correct and ordered fashion. Time regulating federates may send timestamp ordered (TSO) events while time-constrained federates are able to receive TSO events in time order. Each regulating federate must specify a "lookahead" value and ensure that it will not generate any TSO event earlier than its current time plus lookahead [1]. A federate can be both time regulating and time constrained, either of them, or neither. Each federate's LRC maintains two internal event queues, that is, a TSO queue and a FIFO receive queue. When TSO events arrive at a time-constrained federate, the LRC buffers these events in the TSO queue [1]. After requesting a time advance, the federate is passed through all events in the TSO queue with timestamp less than or equal to the federate's granted time. As for the events without timestamp, known as Receive Order (RO) events, the LRC places them in the receive queue in the order in which they arrive. Information in the receive queue is immediately available to the federate.

The RTI services are available as a library (C++ or Java) to the federate developers. Within the RTI library, the class RTIAmbassador [1] bundles the services provided by the RTI. A federate may invoke operations on the interface to request a service (federate-initiated service) from the RTI. The FederateAmbassador [1] is a pure virtual class that identifies the "callback" functions each federate is obliged to provide (RTI-initiated) service. The federate developers need to implement the FederateAmbassador. The callback functions provide a mechanism for the RTI to invoke operations and communicate back to the federate.

The HLA specification leaves the RTI implementation details to the RTI implementers while defining a standard interface. Nowadays there are various RTI software available to the developers, including both commercial and academic implementations. These include the RTI Next Generation (RTI-NG) implementation sponsored by the DMSO [82], the pRTI developed by Pitch AB [2], the Federated Simulations Development Kit developed by the Georgia Tech PADS research group [55], etc. The research work studied in this book is based on the DMSO RTI-NG software; however, the implementation need not be specific to this software. The standard HLA interface enables the use of different RTI implementations for the work discussed in this book.

2.2 Cloning and Replication

Cloning and replication have been widely used in software engineering and distributed applications for various objectives, such as improving performance, enhancing system reliability and scalability, etc. In this section, the terms "cloning" and "replication" are interchangeable with respect to the existing work, while in Chapter 6 a distinction is made between them in our theory of distributed simulation cloning.

2.2.1 Cloning in Programming Languages

Methods of procedure cloning have been studied in the domain of programming languages, such as those presented by Cooper et al. and Vahid [29, 109]. Their approaches aim at optimizing code generation for the compiler, by creating clones of procedure bodies. By carefully partitioning the calls, the compiler ensures that each clone inherits an environment that allows for better code optimization.

Object cloning was discussed in [92] and is used for eliminating parametric polymorphism and minimizing code duplication to overcome some typical inferior performance of object-oriented programs. For instance, polymorphic functions do not know the exact types of the data on which they operate, and must use indirection to operate on them. Plevyak[92] proposed specializing a polymorphic class into a set of monomorphic classes (clones of the corresponding class) to improve efficiency.

2.2.2 Data Replication in Distributed Systems

Data replication techniques play an important role in distributed systems. For the concern of performance, data are often replicated to facilitate fast local access. Scalability is about how to address the performance bottleneck in scaling a system. Data replication is also used as a scaling technique by placing data copies close to the process using them. Moreover, Data replication helps to improve the reliability of a distributed system in the sense that the system continues working when some replicas crash and also provides protection against corrupted data [106].

Data replication has been widely applied in distributed database systems for improving both performance and reliability. In geographically distributed database systems, the access of remote data is expensive. Data are often fully or partially replicated to avoid remote access for read transactions [27]. Full replication requires all data objects being replicated to all sites so that each site holds a complete copy of the distributed database. This extreme case of replication has been recognized not to be the optimal configuration for many applications. Partial replication can consist of replicating (1) all data objects to some sites, (2) some data objects to all sites, and (3) some data objects to some sites [89]. The several partial replication approaches are suggested to balance the overhead in keeping data consistency and the benefit gained in access efficiency. In addition, Data replication has also been exploited to control concurrency in distributed database systems [27].

The general principles and issues studied in Data replication are also used as guidelines in designing and developing our distributed simulation cloning technology. For example, we need to ensure the scalability of our cloning-enabled systems and keep state consistency without sacrificing execution efficiency.

2.2.3 Agent Cloning in Multi-Agent Systems

Multi-agent systems (MASs) are increasingly used when a problem domain is particularly complex, large, or unpredictable [103]. Basically, in MASs an agent has incomplete information or capabilities for addressing the whole problem, and multiple agents interact and collaborate to solve the problem. Shehory et al. use agent cloning as a comprehensive approach to the performance bottlenecks in MASs. Their approach allows agents to make clones, transfer tasks, or merge with other agents [99].

They use an agent cloning approach to deal with overload in MASs. In the context of their approach, agent overload means either an agent is not able to process current tasks due to its limited capability or machine overloads. The first case does not imply machine overload, and therefore they suggest cloning the agent locally to let clones take over part of the tasks. For the second case, their approach will create and activate clones of agents at a remote machine. Through agent cloning, load balancing can be achieved and the performance of an overloaded MAS can be improved.

Agent cloning aims at optimizing the resource usage of the whole MAS. It should be noted that the approach requires an agent to reason about its current and future load. Furthermore, their design and implementation are built upon the RETSINA infrastructure, which is specially designed for intelligent network agents [104].

2.2.4 Object Replication in Parallel Object-Oriented Environments

Concurrent and parallel object-oriented systems have been designed to allow the developers to manage the high complexity of parallel programming and provide modularity and code reuse. The capability of supporting load balancing is one of the major factors affecting the performance of these systems. Jie et al. developed a Parallel Object-oriented Environment for Multi-computer Systems (POEMS) to dynamically balance the processing load based on object replication; thus the performance of parallel object-oriented applications can be improved [70].

Their approach achieves dynamic load balancing by migrating objects. They designed a Parallel Object Replication(POR) model to support task parallelism by distributing the replicas of POR objects in different nodes. Their model only requires migrating the data rather than the entire objects; hence the overheads of migrating objects are reduced considerably.

2.2.5 Fault Tolerance Using Replication

Fault-tolerant techniques often employ redundant/backup components to achieve system robustness. Birman used backup to achieve fault tolerance in building reliable network applications [7]. In [38], fault tolerance was enabled in a distributed system using rollback-recovery and process replication. In rollback-recovery, processes periodically save their states on stable storage during failure- free operation, and the states can be loaded to restore consistent execution. Process replication is used to provide high availability of servers with replicas of server programs running in multiple machines. Principally, these methods take advantage of replication to ensure reliability of distributed applications. Replication consumes extra resources and requires synchronization of the replicas to maintain consistency [34].

2.3 Simulation Cloning

Simulation users often execute multiple "replications" of the same simulation to obtain outputs that meet required confidence levels [74]. In later chapters, the term *replication* is slightly different from that used in traditional simulations, and does not merely mean repeating multiple runs of an application or simulation. In this section, cloning and replication denote making replicas of simulation processes, threads, or data in the middle of execution. We give a survey on existing techniques exploiting cloning/replication in simulations.

2.3.1 Cloning in Rare Event Simulations

The performance of computer and communication systems is commonly characterized by the occurrence of rare events, the probability of cell loss in asynchronous transfer mode (ATM) switches is typically less than 10^{-9}. Because straightforward simulation of rare events often takes an extremely long execution time, rare event simulations are often used to evaluate the performance of this kind of system [50].

Glasserman et al. adopted the cloning approach to improve the efficiency and effectiveness of rare event simulations [48]. In the standard simulation of a stochastic process, a lot of time is spent in the region of state space faraway from the rare set of interest, from where the chance of entering the rare set is extremely low. In order to tackle this problem, their approach splits multiple identical copies of the process to get more chances for rare events to occur when the state space is close to the rare set. Thus, within a given amount of execution time, the approach increases the number of times of hitting the desired rare set. Their approach has a common point with ours in the sense of using cloning to improve execution efficiency. However, our approach is significantly different from theirs. First of all, their approach makes multiple identical copies of the stochastic process while our approach makes clones of a federate with each clone exploring different execution paths. More importantly, our approach aims at distributed simulation cloning, and their approach does not concern distributed simulation at all.

Addie proposed an approach to emphasize the execution paths of interest in quantum simulations by simulation cloning and killing off the uninteresting paths [4]. His approach is intended to increase the speed and accuracy of estimation of rare events. It uses threads to simulate quantum stochastic processes, and makes multiple clones of selected threads or terminates clones in the middle of a simulation following dedicated rules. The clones evolve independently; thus execution efficiency is gained compared with repeating simulations using multiple parameters. His idea also aims to avoid unnecessary or undesired repetition of executions among different execution paths.

2.3.2 Multitrajectory Simulations

Gilmer, Jr. and Sullivan developed a multitrajectory simulation technique to obtain a better understanding of the possible outcome set of stochastic simulation [45, 46]. In conventional stochastic simulation, each replication gives only one outcome. In contrast, the multitrajectory simulation generates multiple outcomes when a random event occurs, with each outcome constituting a trajectory maintaining its own states cloned from the original execution. Their technique has advantages over traditional stochastic simulation and requires many fewer runs to generate an equivalent quality histogram. Multitrajectory simulation technology is also adopted to support recursive simulation, which means a simulation recursively uses the outputs of another instance of the same simulation, thus improving the quality of decision making [47].

In [46], the same authors briefly describe the issue of multitrajectory routes and trajectory management. Their discussion approximates to the rudiments of our scenario management mechanism (see Section 6.4 and Chapter 8). However, their cloning technique often requires the modeler to take care of state cloning and initiating alternative trajectories when building the models, which becomes even more difficult when the stochastic event is embedded in the functional code [45]. Hence, there are limitations to the application of this technique to existing simulation models. Furthermore, their cloning approach does not perform well compared with normal runs when dealing with small scenarios.

2.3.3 Cloning in Simulation Software Packages

Some simulation software packages also incorporate cloning functionality. For example, the Simulation Language with Extensibility (SLX) supports cloning of transactions in its simulation language [52, 53]. SIMPHONY also provides methods for cloning entities in construction simulation [3]. Developers may take advantage of the functionalities provided by those packages to customize their simulation cloning mechanisms.

2.3.4 Parallel Simulation Cloning

Hybinette and Fujimoto were the first to employ simulation cloning technology as a concurrent evaluation mechanism in the context of parallel simulation [60, 61, 62]. Their work provides a thorough study of parallel simulation cloning technology. The motivation for this technique was to develop a cloning mechanism that supports an efficient, simple, and effective way to evaluate and compare alternate scenarios. The method was targeted for parallel discrete event simulators that provide the simulation application developer with a logical process (LP) execution model. Through their work, the basic theory of parallel simulation cloning has been established.

Parallel simulation cloning improves performance of simulation execution in two ways: (1) cloned simulations share the same execution path before the point that cloning happens, and (2) the idea of a virtual logical process avoids repeating unnecessary computation among clones and aims at further sharing computations after the decision point.

Cloning of an LP may occur at predefined Decision points so the original LP and the clones execute along different execution paths in parallel; thus the execution path before cloning is shared. In designing the cloning mechanism, they proposed replicating LPs incrementally rather than copying the whole simulation at once as it is likely that only some of the LPs are different among execution paths. They designed an LP as a combination of a virtual LP and an underlying physical LP. The physical LPs realize the real parallel simulation system and maintain the state space and physical messages. In contrast, a virtual LP only maps to the corresponding physical LP and interacts with other virtual LPs via its physical LP. When cloning a simulation, their approach only requires replicating the physical LPs that are different in the alternative scenarios. For identical LPs, only the virtual LPs need to be replicated (which is cheap) and multiple replicas share a single physical LP. Besides process cloning, their cloning mechanism also includes message cloning. A message is cloned when a shared physical LP sends a message to LPs that are not shared.

Some concepts and issues from parallel simulation cloning are also examined in our research, such as decision point, execution paths and evaluation of alternative paths, etc. Their work gives us a viable guideline in identifying some basic research issues.

Their cloning mechanism, which relies on theGeorgia Tech Time Warp simulation executive [35], is implemented as a Clone-Sim software package. The package requires the simulation executive to support copying of LPs as well as dynamic creation, allocation, and initialization of LPs [61]. Hence, their approach does not imply a generic LP cloning mechanism. Futhermore, their approach is aimed at the parallel simulation paradigm, and it seems that there exists a gap to directly apply their approaches to HLA-based distributed simulation. For example, their design does not concern maintaining state consistency of LPs in the distributed environment on cloning, which is crucial in distributed simulation cloning.

2.3.5 Cloning of HLA-Compliant Federates

Schulze et al. extended the simulation cloning technology to HLA-compliant simulations [96]. They introduced a cloning approach to extend the flexibility of system composition to runtime, and gave a proposal for cloning federates and the federation at runtime [96, 97].

Internal cloning and external cloning techniques were suggested to clone the federates. Internal cloning means replicating federates and letting the replicas participate in the original federation. This technique requires the federates to distinguish the messages from clones. It is also suitable for federates with only passive behaviors, such as a pure "observer" federate, which does not generate any event, or in other words does not affect the states of any other federate during the simulation. The external cloning approach makes replicas of federates so that they operate in a different federation. Forecast functionality is provided by this approach via the parallel scenario management of different time axes.

They describe the problems of cloning federates in terms of dealing with cloning all relevant elements of the simulation model, such as object instances and other internal states of the model. Their work proves the feasibility of applying cloning in HLA-based simulation. However, other critical issues are not clearly addressed, such as how to ensure the correctness of the interaction between the clones (and the federate being cloned) and other related federates. Their cloning design intensively relies on the SLX simulation package, so that their approach lacks generality when applied to other simulation models [52, 53]. The detailed federate replication algorithms and the performance of the proposed approaches are not given.

2.3.6 Fault-Tolerant Distributed Simulation

Distributed simulations are liable to encounter failure, and normally users have to restart the entire simulation session from scratch. Obviously, this results in a considerable waste of time and computation. Approaches for fault-tolerant distributed simulation have been studied to address this problem.

State manipulation methods such as checkpoint and migration are often exploited in existing fault-tolerant approaches for distributed simulations. In [34], a rollback-based optimistic fault-tolerance scheme is integrated with an optimistic distributed simulation scheme. The scheme models a failure as a straggler event to optimize implementation efforts. This kind of approach has some common drawbacks from the developers' perspective:

They require simulation components to have the capability of saving system states and recovering the saved states at the model level in the case of failure. The approach does not apply in the case that existing simulation models are not designed to support state manipulation or process/thread replication.

Berchtold and Hezel proposed the Replica Federate approach for fault-tolerant HLA-based simulation [6]. Their approach produces multiple identical instances of one single federate, and failures can be detected and recovered upon the outputs of those identical instances. However, redundancy is liable to result in lower system performance. Furthermore, extra federate replicas in a distributed simulation increase the probability of overall system failure incurred by an LRC failure.

2.4 Summary of Cloning and Replication Techniques

In this chapter, we introduce the basics on HLA and its supporting software, that is, the RTI. The distributed simulation cloning technology of our study is based on this architecture. Related work on the major research issues in this study, that is, replication and cloning in various computing technologies, distributed systems, and simulations, is described and briefly analyzed. Our research is significantly different from the above research in the following aspects:

- We aim to enable distributed simulation cloning technology based on the High Level Architecture. This is not addressed by the above research except Schulze et al.'s work.

- In our project, simulation cloning means the replication of the entire federate at runtime.

- The proposed cloning technology supports evaluation of concurrent scenarios of a distributed simulation. Automatic scenario partitioning /managing and dynamic execution/computation sharing will be enabled. Our technology is designed to mask the complexity involved in the above efforts.

- As our design targets potential academic or industrial users who may have their own complex simulation models, we have the additional aim to support reusability and transparency while enabling simulation cloning.

- We intend to provide a generic cloning infrastructure to existing distributed simulations, which is independent of the RTI software and the implementation of the simulation model.

- The technology is designed to offer analysts the flexibility to specify the conditions/rules according to which distributed simulation cloning can be initiated automatically.

- The technology ensures the correctness of the synchronization between federates in cloning-enabled simulations.

The proposed solution must guarantee the state consistency of distributed federates. Our technology needs to ensure the scenarios created by cloning have an identical execution to the traditional approach. Designers of replicated systems have to make trade-offs between consistency, performance, and availability [115], and at the same time this technology needs to minimize the extra overhead incurred by cloning. We need to explore the mechanisms of managing a cloning-enabled distributed simulation at both the scenario level and clone (federate) level.

2.5 Fault Tolerance

People have developed many technologies for facilitating fault tolerance in distributed applications. Cristian pointed out some principles about fault tolerance in distributed system architectures [31], and these are understanding failure semantics, masking failure, and balancing design cost.

The checkpoint and message-logging approach is commonly used. For example, as proposed in [71], a process records each message received in a message log while the state of each process is occasionally saved as a checkpoint.

A failed process can be restored using some previous checkpoint of the process and the log of messages. The HLA federation save and restore services [1] could be used to save the RTI states at some checkpoints. In the case of failure, a new federation could be created to "restore" the federation with the saved states. However, in the checkpoint approach, the simulation model should have the functionality to manipulate the states at the model level, and it repeats the computation from one of the checkpoints onward. Moreover, the overhead for executing federation save and restore can be significant [101, 118].

Fault-tolerant techniques often employ redundant/backup components to achieve system robustness. Birman used backup to ensure fault tolerance in building reliable network applications [7]. In [38], fault tolerance was enabled in a distributed system using rollback-recovery and process replication. Principally, these methods take advantage of replication to ensure the reliability of distributed applications. Another typical example is the Replica Federate approach proposed in [6]. This approach produces multiple identical instances of one single federate, and failures can be detected and recovered upon the outputs of those identical instances. However, replication consumes extra resources and requires synchronization of the replicas to maintain consistency; as such, it results in lowered system performance. Furthermore, extra federate replicas in a single federation increase the probability of overall system fault due to an RTI failure, and this may also limit the scalability of the approach. Our fault-tolerant framework adopts both state saving and replication in the design, and it avoids the drawbacks of the above approaches. By separating the execution of simulation models from RTI failures, the framework does not require rollback support from simulation models. The light-weighted physical federation consumes minimal system resource and operates independently from the simulation model execution. The framework makes replicas of physical federates only when failure occurs while leaving simulation models intact. Therefore, the redundancy incurred by common replication approach is also minimized.

Although fault-tolerance support has been informally proposed in the latest HLA Evolved specification and some design patterns for fault-tolerant federations were suggested in [83], there are only a few preliminary and nonstandard implementations for this purpose. In addition, the "failure-over RTI" design pattern suggested in [83] is similar to the scheme proposed by us in [22]. The design pattern provides federates with a prioritized list of RTIs with an active RTI servicing the federates, and they connect to another one when the active RTI fails. Our scheme in [22] suggests replacing a failed RTI with a new one using the decoupled federate architecture.

[36, 37] proposed a framework, Distributed Resource Management System (DRMS), for robust execution of federations. Their framework deals with failure by migrating federates to new hosts upon failure, while using the checkpoint approach for state saving/recovery. In the context of this approach, the federates should be specially developed to save/recover the simulation models' internal states. It is also assumed that federates executed within the scope

of DRMS are portable, meaning that they should not be bound to a specific piece of hardware and can be easily migrated between different host environments. The experimental results indicate a significant overhead for providing fault tolerance in terms of extra messages in some scenarios [36]. In contrast, our framework is not subject to the above constraints. It provides a relatively generic fault-tolerance solution to the HLA-based distributed simulations. However, it is a challenge to develop a generic fault-tolerant model. One of the difficulties is due to the assumption that developers can model their federates in a totally free manner. It is unlikely that a generic state saving and replication mechanism can be provided that will be suitable for any federate. Even given such a mechanism, it is unlikely that all developers will use the same standard package to model their simulations. Without the ability to customize the user's simulation code, it is almost impossible to make snapshots of all system states of any federate. The principle of reusing existing federate code increases the difficulty of this task. On the other hand, the HLA standard makes it relatively easy to intercept the system states at the RTI level using a middleware approach. Furthermore, we can see that the simulation model and the Local RTI Component have very different characteristics. Therefore, it suggests a distinction should be made between these two modules when dealing with failure. It is desirable to develop a generic framework to handle the failures of the RTI rather than the faults of the simulation models.

2.6　Time Management Mechanisms for Federation Community

There only exist few noticeable works on a Time Management mechanism aimed at HLA federation community networks. Lake introduced a preliminary work on Time Management for federations linked by a bridge [73]. The Time Management algorithm is implemented using extensions to the standard HLA specifications. His study does not support a hierarchical federation community. Although a robust Time Management algorithm has been claimed in [73], the algorithm may not always work correctly as it solely relies on the Lower Bound Time Stamp (LBTS) for calculation. The problem is further elaborated in Section 6.2. The system performance has not been presented. Instead of using the RTI communication backbone, the algorithm requires the bridge to broadcast the information of LBTS calculation to all federates per time advance. The effect of this costly operation on system performance or whether such operation can be optimized has not been analyzed.

In [30], Cramp and Best proposed "Distributed Federate Proxy (DFP) Architecture" for interconnecting federations into hierarchical federation communities. They discussed the necessary constraints on the components of a hierarchical federation community in order to ensure correct temporal and causal ordering of events. They claimed that the entire DFP architecture can perform correct time management. However, the correctness of their algorithm has not been mathematically proven, and there are no experimental results for the evaluation of the architecture or validation of the proposed algorithms either. Furthermore, in their design, a special federate called "Proxy Component" corresponds to each federation, and the Proxy Components are interconnected via an additional tree of processes handling inter-federation communications, etc. The inevitable overhead incurred in such design has not been addressed. Whether their approach supports federation communities of other architectures remains unclear.

Another interesting work is the proximity-aware synchronization developed by Okutanoglu and Bozyigit [90]. Their basic idea is to relate the Data distribution in the federation community to the synchronization strategy. In the context of their method, federates are clustered, and synchronization will be waived between two clusters without interactions. However, the method fully bases on the Data distribution Management services and demands that the federates to use these services. Another requirement is the prediction of the interaction pattern among clusters. The features make the method unlikely to be a generic one. Another issue is that the method may encounter a severe problem when two federates do not interact directly but through intermediate federates instead. For example, an agent federate and a factory federate in a supply chain simulation may not exchange messages and they may adopt different regions according to the locations where the two entities persist in the realistic world, but synchronizing the two federates is mandatory as far as causality is concerned.

Part II

Middleware and Software Architectures

3

A Decoupled Federate Architecture

CONTENTS

This chapter introduces the idea of decoupling the Local RTI Component from a normal HLA federate, to give a Decoupled Federate Architecture. This architecture forms the basis for enabling federate cloning at runtime.

3.1 Problem Statement

A normal simulation federate can be viewed as an integrated program consisting of a simulation model and Local RTI Component (LRC), as shown in Figure 3.1. The simulation model executes the representation of the system of interest, whereas the LRC services it by interacting and synchronizing with other federates. In a sense, the simulation model performs local computing while the LRC carries out distributed computing for the model.

Cloning of a federate occurs at a decision point to enable different candidate actions to be performed. "Cloning" implies that the new clones of one particular federate should initially have the same features and states as the original federate, both at the RTI level and at the simulation model level. This is to ensure the consistency of the simulation state. For example, at the RTI level, clones must have subscribed to the same object classes and registered the same object instances, etc. At the simulation model level, the clones

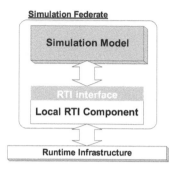

FIGURE 3.1
Abstract model of a simulation federate.

should have the same program structure, data structures, objects, and variables; all these program entities should have identical states. Immediately after the cloning, the clones will be given some particular parameters or routines to execute in different paths.

One possible solution is to introduce a state saving and replication mechanism to the simulation federates, allowing the simulation federate to store snapshots of all the system states. When cloning occurs, new federate instances are started and initialized with stored states. However, users model their simulations in a totally free manner. It is unlikely that a generic state saving and replication mechanism can be provided that will be suitable for any simulation federate. Besides, as the LRC is not designed to be replicated, direct cloning of a federate can lead to unpredictable and uncontrollable failure at the RTI level.

Even given such a mechanism, it is unlikely that all simulation developers will use the same standard package to model their simulations. Without the ability to customize the user's simulation code, it is almost impossible to make snapshots of all system states of any federate. Furthermore, the principle of reusing existing federate code increases the difficulty of this task. On the other hand, the standard HLA specification makes it relatively easy to intercept the system states at the RTI level. Using a middleware approach, one may save and replicate the RTI states while enabling transparency. Thus we can see that the simulation model and the LRC have very different characteristics. Therefore, it suggests a distinction should be made between these two modules for cloning a federate.

Decoupling the LRC from the simulation model has further advantages, such as providing a way of integrating simulation packages or legacy simulations into an HLA-based distributed simulation. Straburger et al. presented different solutions including a gateway program approach to adapt models built using commercial-off-the-shelf (COTS) simulation packages to the HLA

[102]. Under this approach, a simulation application consists of a COTS-based model and a gateway program that couples the model with the RTI. Thus, previously stand-alone models may interoperate with each other using the HLA specification with user transparency maintained. Similarly, McLean and Riddick designed a Distributed Manufacturing Adapter to integrate legacy simulations with the HLA/RTI [80]. This design aims to simplify the integration while gaining the capabilities of HLA-based distributed simulations.

Morse et al. proposed an Extensible Modeling and Simulation Framework (XMSF) to allow various simulations to take advantage of Web-based technologies [85]. They developed a Web-enabled RTI implementation to provide support for HLA-compliant simulations to communicate with the RTI through SOAP and BEEP. Several obvious advantages are identified such as: platform crossing, language crossing, and interoperability of federates (for example, breaking though the constraints of firewalls, etc).

Different from the above work, this project focuses on using a decoupled federate architecture to enable simulation cloning, while facilitating reusability, simulation independency, and user transparency. The design also needs to ensure execution efficiency. The technology may also be used to provide a solution to other issues, such as fault tolerance, supporting a Grid/Web-enabled architecture Web-enabled architecture, and load balancing, as discussed in Chapter 4.

3.2   Virtual Federate and Physical Federate

In the context of the decoupled architecture, a federate's simulation model is decoupled from the Local RTI Component. A virtual federate is built up with the same code as the original federate. As HLA only defines the standard interface of RTI services, we are able to substitute the original RTI software with our customized RTI++ library without altering the semantics of RTI services (see Chapter 7). Figure 3.2(B) gives the abstract model of the virtual federate. Compared with the original federate model illustrated in Figure 3.1, the only difference is in the module below the RTI interface, which remains transparent to the simulation user.

A physical federate is specially designed as shown in Figure 3.2(A). The physical federate associates itself with a real LRC. Physical federates interact with each other via a common RTI. Both virtual federates and physical federates operate as independent processes. Reliable external communication channels, such as Inter-Process Communication (IPC), Socket, SOAP, or other out-of-band communication mechanisms, bridge the two entities into a single federate executive [100]. To ease discussion, this chapter uses the abbreviation "**ExtComm**" to denote those alternative communication channels and assumes messages are delivered / received strictly in an FIFO manner.

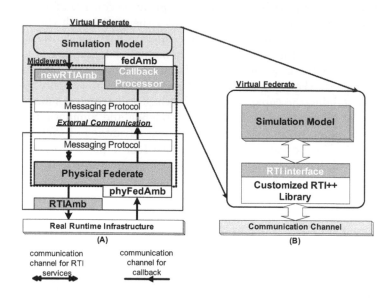

FIGURE 3.2
Decoupled federate architecture.

All the components inside the dashed rectangle in Figure 3.2(A) form a **Middleware** module between the simulation model and the real RTI. Within the virtual federate, the **newRTIAmb** contains customized libraries while presenting the standard RTI services and related helpers to the simulation model. This module can be designed to contain all other management modules for the objectives mentioned previously. The **fedAmb** serves as a common callback to the user federate, which is freely designed by the user and independent of the decoupled approach. The newRTIAmb handles the user's RTI service calls by converting the method together with the associated parameters into ExtComm messages via the **Messaging Protocol**. The protocol mainly defines a mapping between an ExtComm message and the RTI method it represents. For example, an RTI_UPDATE message indicates that the virtual federate has invoked the RTI method *updateAttributeValues()*. The protocol can also be extended for other purposes, such as manipulating the physical federate. The ExtComm conveys these messages immediately to the physical federate for processing in an FIFO manner.

The physical federate converts an RTI call message generated from the virtual federate into the corresponding RTI call through its own messaging protocol layer. The **RTIAmb** module executes any RTI service initiated by the simulation model prior to passing the returned value to the ExtComm. The **phyFedAmb** serves as the callback module of the physical federate to respond to the invocation issued by the real RTI. Within the phyFedAmb module, the messaging protocol is employed to pack any callback method with its parameters into ExtComm messages. The ExtComm enqueues the callback message to the **Callback Processor** module at the virtual federate. Through the messaging protocol, the callback processor activates the corresponding fedAmb method implemented by the user. From the simulation users' perspective, a combination of one virtual federate and its corresponding physical federate operates as a simulation federate in the context of the decoupled architecture. The federate combination performs an identical execution to a normal simulation federate with the same code as the virtual federate. In future discussion, we will explicitly use "normal federate" to refer to a traditional federate that directly interacts with the RTI. By default, in the discussion in this chapter a federate contains a virtual federate module and a physical federate module.

3.3 Inside the Decoupled Architecture

As discussed above, the decoupled approach interlinks a virtual federate and the physical federate into a simulator that performs an identical simulation to the corresponding normal federate. This section gives the details of how an RTI service call is executed and the callback is invoked in the Decoupled federate architecture.

Figure 3.3 depicts the procedure where a simulation model initiates an RTI call and waits for a return from the real RTI, using the *updateAttributeValues* method as an example. The procedure is as follows:

- The virtual federate invokes the redefined *updateAttributeValues* method.

- Inside the *updateAttributeValues* method, the *packMsg* routine extracts the data stored in the *AttributeHandleValuePairSet (AHVPS)* and packs them together with other parameters such as the associated timestamp, object instance handle, and tag into an RTI_UPDATE message.

- The ExtComm enqueues the RTI_UPDATE message to the physical federate. The virtual federate switches to waiting mode for the returned message.

- Once the physical federate receives the ExtComm message, it invokes the *unpackMsg* routine to process it according to the associated type, RTI_UPDATE.

- A new *AHVPS* object and related parameters are recovered based on the ExtComm message and passed to the *RTI::updateAttributeValues*, which invokes the real RTI service.

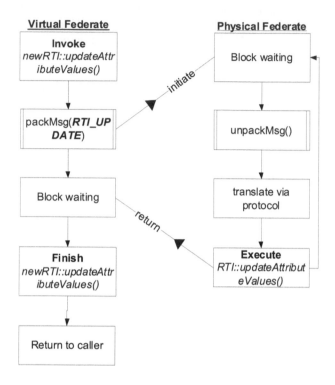

FIGURE 3.3
Executing an RTI call in the decoupled architecture.

- On the accomplishment of this *RTI::updateAttributeValues* call, the physical federate acknowledges the virtual federate with an ExtComm message containing the returned value.

- The *updateAttributeValues* call finishes and the data retrieved from the acknowledgment message is returned to the simulation model.

From the user's point of view, the initiation and accomplishment of an RTI call are identical to the original normal federate. The semantics of RTI services are kept intact in the decoupled approach.

The RTI software has an interface that provides flexible methods to the user for packing update data and leaves the transmission details transparent. The user can create update data of variable lengths. However, most low-level communications do not offer flexible methods to handle message delivery without agreement of lengths between source and destinations. For example, most IPC mechanisms have limitations in message size and buffer size. The **Message Queue** based on Solaris defines the maximum queue length as 4096 bytes [100]. The message sender and receiver must agree with each other on the same message length. If a fixed message size is defined for ExtComm messaging, it may incur some unnecessary overhead. A fixed large size is inefficient in transmitting small messages. On the other hand, a fixed small size increases the overhead for packing, delivering, and unpacking a large number of small packets in the case of processing large messages. Therefore a simple protocol is proposed for messaging between the virtual federate and physical federate. We define a small message size (MSG_DEF) and a large message size (MSG_LG) for assembling user data into packets. A special "PREDEFINED" packet is used to notify the receiver if large or multiple packets are to be sent for a single data block. Figure 3.4 gives the messaging details based on this simple protocol.

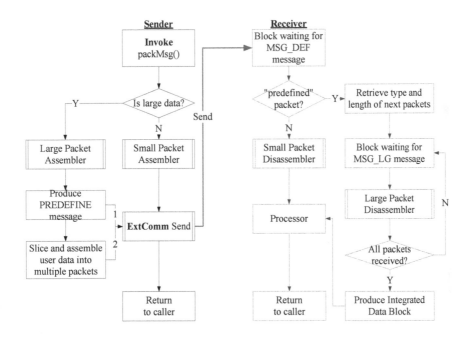

FIGURE 3.4
Messaging between virtual federate and physical federate

The RTI communicates with a federate via its Federate Ambassador provided by the user [82]. In the DMSO RTI, a federate must explicitly pass control to the RTI by invoking the *tick* method. For example, the RTI delivers the Timestamp Order (TSO) events and Time Advance Granted (TAG) to a time-constrained federate in strict order of simulation time, which coordinates event interchange among time-regulating and time-constrained federates in a correct and causal manner. Therefore, the decoupled architecture should guarantee that (1) the Federate Ambassador at the user federate works in a callback-like manner and (2) callback methods are invoked in the correct or-

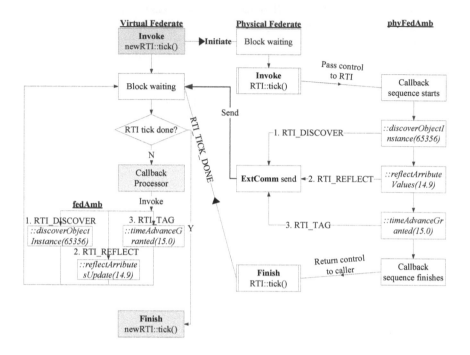

FIGURE 3.5
Conveying callbacks to the virtual federate.

der. Figure 3.5 depicts how to realize these functionalities. To ease discussion, we assume the physical federate will get the callbacks shown in Figure 3.5. This procedure is illustrated by the following steps:

- The virtual federate invokes the routine newRTI::tick and the latter sends out an RTI_TICK message to the physical federate.

- The physical federate calls the real RTI *tick* according to the RTI_TICK message.

- The LRC acquires control and delivers events to the phyfedAmb module of the physical federate in a strict order.

- In each callback method invoked, the data sent by the RTI is enqueued to the callback ExtComm channel. The routine inside the *newRTI::tick* accesses the queue for the virtual federate.

- As long as the RTI_TICK_DONE message is not detected, the callback processor continuously processes the messages in FIFO order while activating the corresponding method in the fedAmb module based on the messaging protocol.

- At the physical federate side, once the RTI finishes its current work and passes control to the physical federate, the latter returns an RTI_TICK_DONE message to the virtual federate.

- On receiving the RTI_TICK_DONE message, the virtual federate accomplishes the *newRTI::tick*, and control is returned to the caller immediately.

The real RTI starts to take charge only when the physical federate explicitly invokes *RTI::tick*. On the other hand, the *newRTI::tick* can only return when the real RTI finishes its work. As the ExtComm channels work in an FIFO manner, the order of each callback method invoked at the physical federate is identical to the sequence in which the callback processor at the virtual federate processes the data. From the user's perspective, the callback mechanism based on the decoupled approach executes the equivalent operations to the normal federate. It guarantees consistency in presenting messages from the real RTI to the simulation model and also ensures user transparency.

The decoupled architecture requires an additional ExtComm communication layer although it performs exactly the same computation as the corresponding normal federate. The external communication may incur some extra overhead. To investigate the overhead incurred by the decoupled approach, a series of benchmark experiments has been performed to compare with the normal federates. Section 3.5 reports and analyzes the experimental results in terms of event transmission latency and synchronization efficiency.

3.4 Federate Cloning Procedure

Using the decoupled approach can solve the problem caused by making copies of the LRC at runtime in the case of executing traditional federates. Cloning a federate can be achieved by replicating the virtual federate process and starting an additional physical federate instance with restored system state.

As illustrated in Figure 3.6, at runtime the middleware intercepts the invocation of each RTI service method. The interceptor logs all the RTI system states into stable storage. Some RTI states are relatively static, such as the federate identity, federation information, the published/subscribed classes and time constrained/regulating status. Other states include the registered or deleted objected instances, and granted federate time. Some event data may also need to be saved, such as sent and received interactions, updated and reflected attribute values of object instances, etc.

As soon as a federate reaches a decision point, the cloning conditions are checked to decide whether or not cloning is required. If necessary, the middleware spawns clones of the federate immediately to explore alterative execution paths. From the perspective of the federate making clones, the simulation cloning procedure can be described as follows (see Figure 3.7):

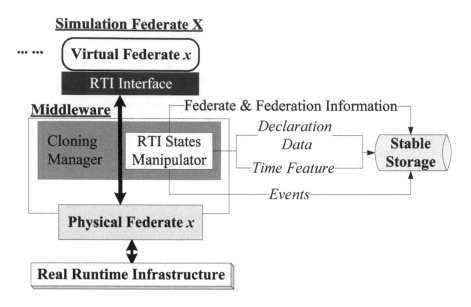

FIGURE 3.6
Saving RTI states.

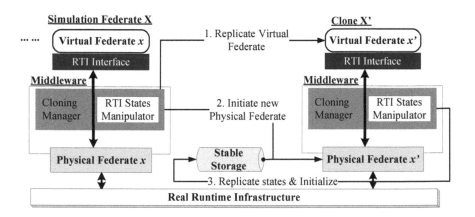

FIGURE 3.7
Federate cloning.

- **Copying simulation model**. The cloning manager within the middleware makes the specified number of clones of the virtual federate (simulation model).

- **Initiating physical federates.** The cloning manager initiates an individual physical federate for each clone of the virtual federate and hooks up the two new processes into a whole.

- **Replicating states.** A clone's physical federate is initialized with the stored system states from the parent federate, after which a new clone of the original federate is formed.

The distributed simulation cloning algorithms are described in detail in Chapter 9.

3.5   Benchmark Experiments and Results

The decoupled architecture separates the simulation model and the LRC into independent processes. In contrast, in a normal federate these operate within the same memory space, which may be superior in terms of execution efficiency. In order to investigate the overhead incurred in the proposed architecture due to decoupling, we perform a series of benchmark experiments to compare the Decoupled federate with a normal federate.

The experiments study the scalability by emulating the simulation cloning procedure based on the decoupled architecture using the IPC *Message Queue* [100] as the external communication to bridge the virtual and physical federate. Through these experiments, we also find out the overhead of adopting the *Message Queue* as the external communication backbone. The performance is compared in terms of latency and time advancement calculation. Latency is reported as the one-way event transmission time between one pair of federates. The time advancement performance is represented as the time advance grant rate.

3.5.1   Experiment Design

The experiments use three computers in total (two workstations and one server), in which the server executes the RTIEXEC and FEDEX processes (see Figure 3.8). The federates that run at one independent workstation are enclosed in a dashed rectangle. In our case, *Fed A[i]* and *Fed B[i]* ($i \geqslant 1$) occupy Workstation 1 and Workstation 2, respectively. The computers are interlinked via a 100Mbps-based backbone. Table 3.1 gives the configuration of the test bed.

The experiments study the scalability by increasing the number of identical federates. As shown in Figure 3.8, *Fed A[1]* and *B[1]* form a pair of initial federate partners, which represent the federates to be cloned in the original scenario. *Fed A[i]* and *B[i]*(i>1) stand for the *i*th clones of the two initial federates respectively and form an independent scenario. The architecture is used through all the benchmarks experiments and for both normal federates and Decoupled federates.

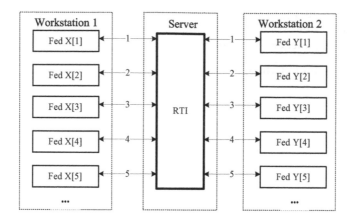

FIGURE 3.8
Architecture of benchmark experiments for decoupled federate architecture

TABLE 3.1
Configuration of Experiment Test Bed for Examining Decoupled Federate Architecture

Specification	Computers		
	Workstation1	Workstation2	Server1
Operating System	Sun Solaris OS 5.8	Sun Solaris OS 5.8	Sun Solaris OS 5.8
CPU	Sparcv9 CPU, at 900 MHz	Sparcv9 CPU, at 900 MHz	Sparcv9 CPU * 6, at 248 MHz
RAM	1024M	1024M	2048M
Compiler	GCC 2.95.3	GCC 2.95.3	GCC 2.95.3
Underlying RT	DMSO NG 1.3 V6	DMSO NG 1.3 V6	DMSO NG 1.3 V6
Processes running on	Fed A[i]	Fed B[i]	RTIEXEC &FEDEXE

A DDM-based approach is used to partition concurrent scenarios (see Chapters 7 and 8). For the latency benchmark, each pair of federates has an exclusive point region associated to any event being exchanged. The federates are neither time regulating nor time constrained. In one run, each federate updates an attribute instance and waits for an acknowledgment from its partner (from *Fed A[i]* to *Fed B[i]*, and vice versa) for 5,000 times with a **payload of**

100, 1k, and 10k bytes. The time interval in the ping-pong procedure will be averaged and divided by 2 to give the latency in **milliseconds**. A federate merely reflects the events with identical region to itself. In other words, *Fed A[i]* only exchanges events with *Fed B[i]*.

As for the time advancement benchmark, all federates are time regulating and time constrained. Each federate has lookahead 1.0 and advances the federate time from 0.0 to 5,000.0 with timestep 1.0 using *timeAdvanceRequest* [82]. The results report the rate that the RTI issues *timeAdvanceGranted* (TAGs/second).

3.5.2   Latency Benchmark Results

The latency benchmark experiments report the latency with three different payload sizes. From Figure 3.9 to Figure 3.11, we can see that no matter whether the payload size is small or large, the latency increases steadily with the number of federates. The increment becomes obvious when the number of federates exceeds four pairs (eight federates in total). As indicated in Figure 3.9 and Figure 3.10, when the payload is not greater than 1,000 bytes, the latency varies from about 10 milliseconds for one pair of federates to about 30 milliseconds for seven pairs of federates. The Decoupled federate and normal federate show similar results in this situation, and the Decoupled federates incur only slightly more latency than the normal ones.

As shown in Figure 3.11, when a bulky payload as large as 10,000 bytes is applied, the Decoupled federates incur about 5 milliseconds extra latency to the normal ones. However, the extra latency remains nearly constant with the number of federate pairs. The latencies for both types of federates increase more rapidly than the small payload cases. This is due to the extra overhead incurred by Inter-Process Communication, which becomes obvious with bulky data transmission between the physical federate and virtual federate. When the payload size and the number of participating federates are not too large, the Decoupled federate has a similar performance to the normal federate in terms of latency.

3.5.3   Time Advancement Benchmark Results

Another series of experiments is carried out to compare the Decoupled federates and normal federates in terms of the speed of time synchronization. Figure 3.12 presents the experimental results reported as the TAG rate. In the time advancement benchmark, the TAG rate decreases with the number of federates for both decoupled and normal federates. The rate decreases less rapidly when the number of federate pairs is greater than four (eight federates in total). The TAG rate is about 300 times per second for one pair of federates down to about forty times per second for seven pairs of federates. The results indicate that the performance of the decoupled and normal federates is very similar in terms of time advancement.

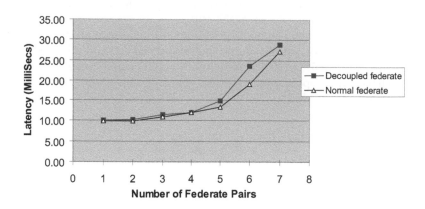

FIGURE 3.9
Latency benchmark on decoupled federate versus normal federate with payload 100 bytes (in milliseconds).

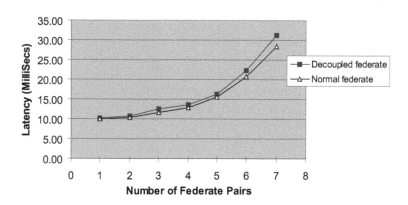

FIGURE 3.10
Latency benchmark on decoupled federate versus normal federate with payload 1,000 bytes (in milliseconds).

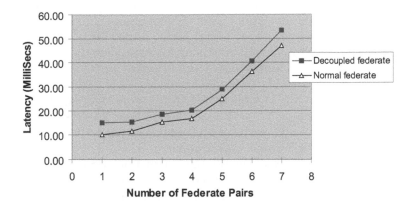

FIGURE 3.11
Latency benchmark on decoupled federate versus normal federate with payload 10,000 bytes (in milliseconds).

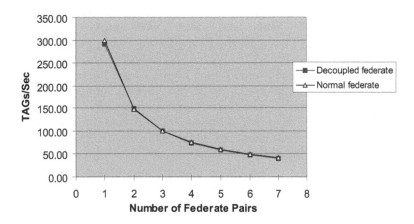

FIGURE 3.12
Time advancement benchmark on decoupled federate versus normal federate (times/sec).

3.6 Exploiting the Decoupled Federate Architecture

In Chapter 3 we discussed the Decoupled federate architecture, which separates a federate's simulation model from the Local RTI Component (LRC).

The architecture enables efficient simulation cloning while facilitating reusability of legacy federate code and user transparency. The Decoupled federate architecture can provide other advantages to distributed simulation technology. Executing HLA-based simulations normally requires users to allocate computing resources to federates and the RTIEXEC (DMSO RTI) statically, which lacks flexibility. The interoperation among federates over the Internet is also restrained by security requirements (e.g., firewalls) or heterogeneity of the computing resources in different organizations.

Web services provide interoperability between software applications distributed over the Internet. As an extension of Web services, the Grid has been designed to manage sharable and often heterogeneous computing resources distributed in different organizations.

One of the advantages of distributing simulations over a network is the capability of exploiting indexsharable computing power. In the case where an individual federate experiences high congestion at the host where it executes, load balancing [119] can improve the overall simulation performance. This can be achieved by migrating load of the congested federate to other lightly loaded hosts [116].

3.6.1　Web/Grid-Enabled Architecture

The Grid infrastructure provides dependable, consistent, pervasive, inexpensive, and secure access to high-end computational capabilities [39]. The Open Grid Services Architecture (OGSA) extends the Web services to include additional functionalities. With these features, Web and Grid services offer a promising approach to enhance the flexibility and interoperability for large-scale HLA-based simulations.

The HLA does not define security mechanisms for using RTI services, while the Grid provides a built-in security architecture to support authentication and secure communication in accessing Grid services [49].

The HLA supports interoperability among HLA-compliant federates. However, this feature is often hampered in the context of large-scale distributed simulations over the Internet. This is due to various security requirements on the networking resources belonging to different organizations. There is a need to combine the HLA and Web or Grid technologies in order to achieve the advantages of both. Directly utilizing Web or Grid services will incur extra effort from the developers in modeling simulations. An alternative approach is to use a Web or Grid Enabled Architecture, which manages the RTI services at the backend while presenting dynamic access of the RTI services to the simulation model or the end users via Web or Grid services. The architecture can be used to relieve the developers from the complication of coding Web or Grid services in existing simulation models.

Web and Grid technologies can help large-scale distributed applications in terms of resource coordination and connectivity. Using a Decoupled federate architecture can ease the combination of Web/Grid technologies with the HLA.

Figure 3.13 gives an abstract model that supports a Web/Grid-Enabled Architecture. The model treats physical federates and the RTIEXEC process as components managed by the **Web/Grid services**. Therefore, each physical federate (or RTIEXEC) is encapsulated as a Web/Grid service and becomes a Web/Grid-enabled component accessible via Web-based communications (such as SOAP, BEEP). A Web/ Grid-enabled RTI library is embedded in the middleware to bridge the model and the physical federate.

The model has significant advantages compared with traditional simulation using the HLA RTI in the sense that (1) it can cross firewalls, (2) it enhances the interoperability and connectivity among simulation federates, and (3) it eases the interoperation between HLA federates and non-HLA simulations. The communication between the virtual federate and the physical federate is via Web services that can pass through most firewalls. Using Web services allows interoperation of federates physically located at various organizations, while connecting them over a normal RTI may encounter network barriers. The RTI component becomes a Web service in the context of the architecture. Non-HLA simulations can also be wrapped as Web services. Therefore, HLA simulations can easily interoperate with non-HLA ones through Web services.

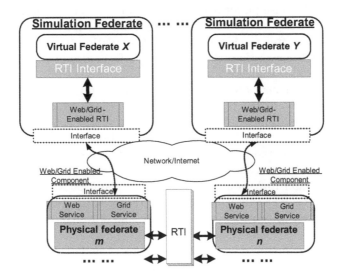

FIGURE 3.13
Model supporting Web/Grid-Enabled Architecture.

Using Grid services has the added benefits over using pure Web services as follows:

- Grid services provide dynamic creation and life-cycle management. The

RTI and federate instances are dynamic and transient service instances under the management of the Grid service infrastructure.

- The model also eases the discovery of the Grid-enabled RTI and federate services using the Grid Index Service [49].

- The Grid provides security in accessing services using its built-in mechanisms such as authentication, secured communication, etc.

- With the notifications feature of the Grid, the simulation model does not have to block for the completion of RTI calls and may operate asynchronously from the physical federate.

With the nature of the Grid, this model has the potential of providing fault tolerance and load balancing. The model supports the reusability of legacy federate code and it also allows federates developed upon the Web/Grid-Enabled Architecture to interact with other traditional federates.

3.6.2   Load Balancing

A large-scale distributed simulation running over the Internet is liable to encounter congestion in some computing hosts or over part of the network. The congestion will cause performance degradation of some federate(s); thus the execution efficiency of the whole federation may be lowered. Load balancing technologies are often used in a distributed simulation (system) to smooth out periods of high congestion; therefore the performance of the system can be improved.

Load management of federates at the simulation model level has been addressed by some researchers, such as the approach proposed in [11]. However, there exists significant overhead and implementation complexity for dealing with RTI-related computing during the migration of a federate at runtime [118]. For example, the federate being migrated needs to resign from the federation and the new federate needs to join the federation. The test bed described in Section 10.2 shows that it can take about 20 seconds for a federate to complete the RTI call *joinFederationExecution* [1]. The migration design also needs to ensure correct event delivery and state consistency at the RTI level.

This section presents a generic model to provide load balancing in the HLA. Our approach treats the simulation model and the physical federate as separate tasks to be managed, and attempts to distribute the model's load to lightly congested nodes when necessary. Appropriate existing algorithms for load management may be adopted. The approach focuses on avoiding the complication of dealing with RTI related operations while migrating running federates.

An abstract model for load balancing is illustrated in Figure 3.14, in which the Resource Observer monitors the availability of resources. In the case of host congestion, the model may alleviate the lack of computing resources by

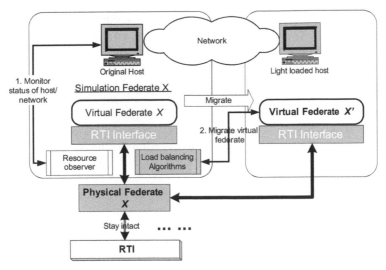

FIGURE 3.14
Load balancing model for simulation model.

migrating the virtual federate to another idle host. The migration process is managed based on the load balancing algorithms. The migrated virtual federate (a replica of the old one) on the new host links to the original physical federate, and both processes form a federate executive to continue the simulation session.

The load balancing model separates the simulation model's migration from the physical federate. The model has advantages in that (1) it does not require the physical federate to perform extra RTI operations to support the model migration, (2) it avoids dealing with in-transit events during migration, (3) it saves the overhead for joining the federation, and (4) the physical federate remains intact while servicing the new virtual federate as before.

3.7 Summary

In this chapter we have investigated the issues related to a Decoupled federate architecture in large-scale HLA-based distributed simulations. A federate is separated into a virtual federate process and a physical federate process, where the former executes the simulation model and the latter provides RTI services at the backend. A standard RTI interface is presented to support user transparency, while the original RTI component is substituted with a customized library. The architecture enables a relatively generic method of replicating the simulation model and also facilitates state saving and replication at the RTI level. The proposed approach guarantees the correctness of executing RTI services calls and reflecting RTI callbacks to the simulation model.

Benchmark experiments have been performed to investigate the overhead incurred by a architecture. The experimental results are compared for a Decoupled federate and normal federate in terms of latency and time advancement performance. The results indicate that the decoupled architecture incurs only a slight extra latency in the case of a bulky payload and has a very close performance of time advancement compared with a normal federate. The results also indicate that an IPC mechanism, such as the *Message Queue*, can provide efficient communication that bridges the virtual and physical federates and is appropriate in designing a framework for federate cloning.

4

Fault-Tolerant HLA-Based Distributed Simulations

CONTENTS

Large-scale High Level Architecture (HLA)-based simulations are built to study complex problems, and they often involve a large number of federates and vast computing resources. Simulation federates running at different locations are liable to failure. The failure of one federate can lead to the crash of the overall simulation execution. Such risk increases with the scale of a distributed simulation. Hence, fault tolerance is required to support runtime robustness. This chapter introduces a framework for robust HLA-based distributed simulations using a "Decoupled federate Architecture." The framework provides a generic fault-tolerant model that deals with failure with a "dynamic substitution" approach. A sender-based method is designed to ensure reliable in-transit message delivery, which is coupled with a novel algorithm to perform effective fossil collection. The fault-tolerant model also avoids any unnecessary repeated computation when handling failure. Using a middleware approach, the framework supports reusability of legacy federate code, and it is platform neutral and independent of federate modeling approaches. Experiments have been carried out to validate and benchmark the fault-tolerant federates using an example of a supply-chain simulation. The experimental results show that the framework provides correct failure recovery. The results also indicate that the framework only incurs minimal overhead for facilitating fault tolerance and has a promising scalability.

4.1   Introduction

Distributed simulation technology facilitates the construction of a large-scale simulation with component models that can be developed independently on heterogeneous platforms and distributed geographically. The High Level Architecture (HLA) defines the rules, interface specification, and object model template to support reusability and interoperability among the simulation components, known as federates. The Runtime Infrastructure (RTI) software supports and synchronizes the interactions among different federates conforming to the HLA standard [33] to give an overall simulation application, known as a federation.

In the case where the problem domain is particularly complex or involves multiple collaborative parties, analysts often need to construct a large-scale federation with individual simulation federates interacting over the Internet. Some typical examples are military commission rehearsal, Internet gaming, supply chain simulation, etc. Those applications usually are time consuming, computation intensive, and require vast distributed computing resources. Simulation federates running at different locations are liable to failure: as the current IEEE 1516 HLA standard does not support a formal fault-tolerant model [65], the crash of a federate or a part of a federation may lead to the failure of the whole federation. When failure occurs, even it is feasible to restart the simulation from a previous checkpoint [71], repeating the execution could either be costly or result in the loss of functions of the failed simulation (e.g., a random event may not be regenerated in the new "recovered" simulation execution). The risk of such failure increases with the number of federates inside a single federation. Hence, there exists a pressing need for a mechanism to support runtime robustness in HLA-based distributed simulations.

A normal federate usually exists as a single process at runtime, and the simulation model shares the same memory space with the Local RTI Component (LRC) [1, 72]. In the case where the RTI crashes or meets congestion, the failure of any LRC prevents the simulation execution from proceeding correctly even though the simulation model contains no error at all. Thus, providing fault tolerance to federates requires an approach to "isolate" the error of the LRC from the simulation model in addition to the challenge of developing a generic state saving and recovery mechanism. In [22], we have proposed a "Decoupled federate Architecture" approach to enable state saving and recovery for federate cloning, and we have also suggested a preliminary scheme to achieve fault tolerance using the architecture [17]. In this chapter we focus on the investigation of the fault-tolerance issue, and whether the Decoupled federate Architecture can be used for this purpose.

This study aims to explore a solution to runtime robustness upon existing RTI implementations, and it also provides the designers of the future RTI software with a viable direction to address the fault-tolerance issues. Our framework has been designed with the following objectives and scope:

- Tackling unpredictable failure of RTI services, regardless of what causes the failure

- Minimizing overhead for providing runtime robustness to ensure execution efficiency

- Resuming normal execution exactly from where a failure occurs without repeating or disrupting the global simulation execution

- Providing user transparency, which (1) avoids the need for developers to include extra "fault-tolerant codes" in modeling federates, to minimize development cost and support reuse of legacy federates; (2) allows developers to model their federates freely using various software packages and on different platforms; (3) masks failure from the users at runtime; and (4) allows users to deploy/execute fault-tolerant federations in the same way as normal federations

This chapter proposes a framework that takes advantage of the Decoupled federate Architecture to handle an RTI failure. The basic idea is to prevent a local failure from affecting the overall distributed computation (simulation). A generic fault-tolerant model has been developed as middleware transparent to the user. The model dynamically substitutes the crashed RTI components with backups, while the simulation federates still continue to operate as normal without being disrupted. The fault-tolerant model avoids repeating the execution of federates when handling failure. Furthermore, the framework uses a sender-based method to ensure reliable in-transit message delivery in case of failure. We have also designed a novel algorithm to dispose of buffered events after they have been successfully delivered to the subscribers. A series of experiments has been performed to validate and benchmark the fault-tolerant model.

The remainder of this chapter is organized as follows: Section 4.2 gives an overview of the Decoupled federate Architecture. Section 4.3 details the functionalities and design of the framework, as well as the algorithms for dealing with in-transit messages. Section 4.4 presents the experiments based on a distributed supply-chain simulation example, which examine the correctness of the fault-tolerant model and compare the robust federates with normal federates in terms of execution efficiency and scalability. In Section 4.5, we conclude with a summary and proposals for future work.

4.2   Decoupled Federate Architecture

As shown in Figure 4.1(A), in an HLA-based distributed simulation, a normal simulation federate can be viewed as an integrated program consisting of a simulation model and Local RTI Component (LRC) [1]. The simulation model executes the representation of the system being analyzed, whereas the LRC services it by interacting and synchronizing with other federates. In a sense, the simulation model performs local computing while the LRC carries out distributed computing for the model.

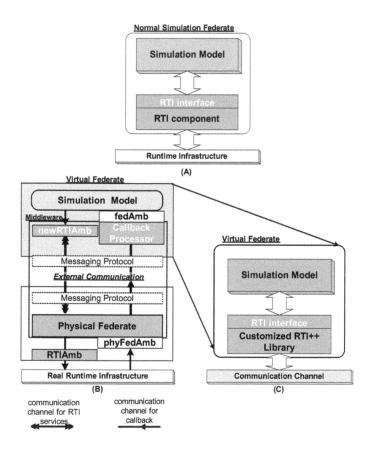

FIGURE 4.1
Normal federate and decoupled federate architecture.

The Decoupled federate Architecture [22] was initially designed to tackle the problems involved in replicating running federates for distributed simulation cloning. It separates a federate's simulation model from the Local RTI

Component. A virtual federate is built with the same code as the original federate. Figure 4.1(C) gives the abstract model of the virtual federate. Compared with the original federate, the only difference is in the module below the RTI interface, which remains transparent to the users. A physical federate (PhyFed) is designed as shown in Figure 4.1(B), and it associates itself with a real LRC. Physical federates interact with each other via a common RTI and form a "physical federation" serving the overall simulation. Both the virtual federate and physical federate operate as independent processes. Reliable external communication channels link the two modules into a single federate executive. The virtual federate and the physical federate may operate within the same address space or in different machines in a networking environment, depending on the developers' requirements.

A well-designed Decoupled federate Architecture can provide federated simulations with almost equivalent execution efficiency to that obtained using normal federates in terms of both latency and time advancement performance [19, 22]. As the Decoupled federate Architecture keeps the standard HLA interface, we can customize our own library (middleware) to expand the functionalities of the original RTI software without altering the semantics of RTI services. With these merits, the architecture seems to be an infrastructure suitable for developing the fault-tolerant model (see Section 7.1).

4.3 A Framework for Supporting Robust HLA-Based Simulations

This section introduces the internal design of the fault-tolerant model and related issues. No implementation can ensure that any program is immune from all faults, and the focus of this study is to develop a robust infrastructure for facilitating distributed simulations rather than to free developers from validating their simulation models. Therefore, the fault-tolerant model does not consider federate crashes due to the incorrect implementation of its simulation model or address deadlock in federation synchronization. It assumes also that (1) the underlying RTI software is properly implemented, which improbably contains bugs; and (2) the messages sent and received in the network are not corrupted as well. An RTI failure in the current implementation can be (1) time-out of an RTI invocation, (2) a critical RTI exception[1], (3) any other unknown error[2] from the RTI, or (4) crash of the physical federate or RTIEXEC/FEDEXEC. Apparently, crash of RTIEXEC/FEDEXEC only concerns the DMSO RTI software. In the remainder of this chapter, a federate means one that contains a virtual federate and a physical federate, and we

---

[1] RTI::RTIinternalError or any other exception specified as critical by the user.

[2] Some low level error that cannot be caught as RTI exceptions defined in the HLA, for example, a runtime error of some system libraries employed by the RTI software.

explicitly refer to a traditional federate that directly interacts with the real RTI as a "normal federate."

4.3.1   Fault-Tolerant Model

In the framework, the fault-tolerant model is embedded in the customized RTI++ library (middleware of the Decoupled federate Architecture). As shown in Figure 4.2, the model contains a Management Module and a Failure Detector in the middleware. The Management Module comprises an RTI States Manipulator and a Buffer Manager.

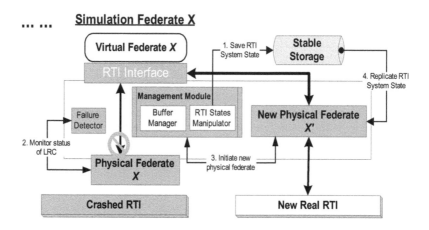

FIGURE 4.2
Fault-tolerant model upon dynamic LRC substitution.

At runtime, the middleware intercepts the invocation of each RTI service method. The RTI States Manipulator saves RTI states immediately before passing the RTI call to the physical federate to execute it. For example, when the virtual federate invokes publishObjectClass, the RTI States Manipulator intercepts this call and saves the information, after which it will call the physical federate via the External Communication channel. In this way, the RTI States Manipulator logs all the RTI system states into local stable storage. Some RTI states are relatively static, such as the federate identity, federation information, the published/subscribed classes, and time constrained/regulating status. Other states include the registered or deleted object instances, and granted federate time. Some event data may also need to be saved, such as sent and received interactions, updated and reflected attribute values of object instances, etc. The RTI States Manipulator logs those states through the standard RTI interface, and its design is transparent to and independent of the underlying RTI implementation. The Buffer Manager makes use of saved attribute updates and interactions for dealing with in-transit events (see Section 7.2 for details). The Failure Detector monitors the status of the LRC or even the RTIEXEC/FEDEXEC if necessary. In the four cases of RTI failure (see above), the first three cases can be detected passively via the physical federate, while the fourth requires the failure detector actively checking the status of the physical federate or RTIEXEC/FEDEXEC. Subsequent to confirming the occurrence of an RTI failure, the Management Module will start a failure recovery procedure. Management Modules of other federates will eventually detect the "remote" failure. In this section we describe a straightforward recovery scheme (as shown in Figure 4.3) from the perspective of the first failed federate(s) using the following steps:

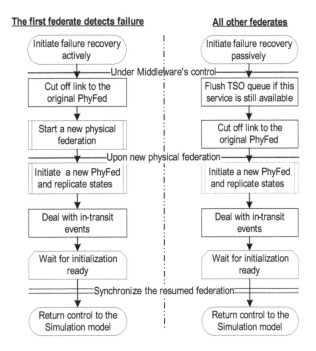

FIGURE 4.3
Illustration of straightforward failure recovery procedure.

- **Preparation for recovery.** The Management Module cuts off the connection from its PhyFed and terminates it, while other federates' middleware attempts to extract received events before doing this.

- **Initiation of new physical federation.** Because the original physical federation cannot function properly due to the RTI failure, the Management Module has to create a new physical federation and initiates a new PhyFed instance. Other federates' middleware also performs exactly the same operation. All virtual federates switch to the new PhyFeds and form a new workable federation together.

- **State recovery.** All RTI States Manipulators recover RTI states from stable storage to the PhyFeds.

- **Handling in-transit events.** All Buffer Managers ensure that in-transit events are delivered properly to the subscribers.

- **Coordination among Management Modules.** The Management Module synchronizes the recovered federation to guarantee that all federates are fully reinitialized and ready to proceed.

Finally, the virtual federates obtain control again and continue execution with the support of a new physical federation. Therefore, physical federates work as plug-and-play components, and they can be replaced at runtime. The fault-tolerant model functions as a firewall to prevent failure of local or remote LRCs from stopping the execution of the simulation model.

4.3.2 Dealing with In-Transit Events

The current design of the fault-tolerant model supports the conservative time synchronization scheme [65]. The RTI does not keep the events, such as updating of attributes and sending interactions, after they have been processed. Thus, the recovered federates may miss some events previously generated with a timestamp greater than the federation time on failure, which should be delivered to them. Examples of this problem are shown in Figure 4.4. Although the example is discussed based on the timestamp ordered (TSO) events, it is similar for receive order (RO) events.

As illustrated in Figure 4.4(A), at simulation time T, Fed[1] sends a TSO event EvX with timestamp $T + \Delta t_1 (\Delta t_1 > \Delta t_0 > Lookahead)$ to subscribers (e.g., Fed[2]). In the case that Fed[2] encounters RTI failure at time $T + \Delta t_0$, Fed[2] resumes with a new PhyFed. But the recovered federate will never receive the event EvX as it has already been lost due to the failure. In another case (Figure 4.4(B)), Fed[3] encounters RTI failure at time T immediately after sending a TSO event EvY with timestamp $T + \Delta t_0$. Refer to the failure recovery procedure shown in Figure 4.3; in this case the message EvY may or may not be received by the Fed[1] before it flushes its original PhyFed's TSO queue to initiate a new PhyFed.

In order to ensure that in-transit events are delivered to the receivers when the simulation resumes from failure, a solution is proposed to resend "image" events with an identical content/timestamp to the corresponding in-transit events generated previously. The Buffer Manager Buffer Manager (Figure 4.2) at the sender side records each outgoing TSO event and indexes the event in time order. The buffer can be flushed to stable storage from time to time. This approach has similarity to the commonly used message logging approach in the sense of recording events, but it does not require rollback of the model's execution [106]. The approach needs to make a trade-off between redundancy in message passing and complexity of the control mechanism under the condition that the new PhyFed must not miss any event that ought to be received. A general principle in designing the resending approach is to ensure that all federates operate in the same way as normal federates that have not encountered a fault.

To minimize extra networking overhead, the proposed approach requires the sender only to resend those events that (1) have been subscribed and (2) have not been received or buffered in the subscriber's TSO queue. The middleware can be designed to help the subscribers notify the particular sender(s) about the reception status of the events originating from the sender. According to the feedback, the sender can selectively generate the required events. The procedure is as follows, including preparation before a crash and the action on failure recovery:

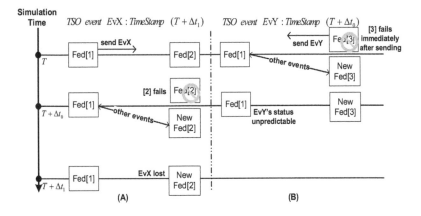

FIGURE 4.4
Illustration of the problems in dealing with in-transit events.

- **Collecting Subscription/Publication/Registration Data.** Each federate builds a Federate Subscription/Publication/Registration (FSPR) Table, which records the classes subscribed/published and the objects registered by other federates. Each federate broadcasts its subscription/publication information that enables other federates to update corresponding entries in their own FSPR Tables. When an object instance[3] is registered (with federate ID encoded using middleware), each subscriber updates the table according to the object class and federate to which it

---

[3]When ownership transferred, the table must be updated accordingly.

belongs. Thus, when attribute updates of an object instance (events) are received, the receiver can trace the source of this event[4].

- **Buffering Events.** Each sender records its local updates in time-stamp order according to their associated object classes. Thus, each sender records what events it has generated. Referring to the FSPR Table, each sender also knows which federates should receive these events.

- **Regenerating Events.** On recovering from failure, the PhyFed being recovered requests the senders to deliver those events with timestamp greater than its current granted time[5]. Thus, the recovered federate can receive those events and pass them to the simulation model with the advance of time.

The resending approach is also applicable for processing interactions. For processing RO events, the Buffer Manager logs and indexes all the outgoing RO events according to the sequence in which they are created; thus the index can be used to identify the RO events. The subscribers simply keep the indexes of the RO events they have received. When failure occurs, the PhyFed being recovered requests the senders to deliver the missing RO events. The RO events to be resent can be easily identified by comparing the indexes maintained by the sender and the receiver. After that, the senders resend the missing RO events and dispose of the logged RO events accordingly after successful delivery. This approach is similar to the counter mechanism introduced in [12]. The Data Distribution Management (DDM) method used in [117] can be also adopted in our case to optimize the delivery of missing events.

4.3.3   Fossil Collection

Using the scheme described in the previous section, sent events are buffered at the sender's side against any potential unpredictable failure. As the simulation execution proceeds, the buffered data will accumulate indefinitely; at some stage this will become a bottleneck as system resources are wasted in maintaining a huge amount of redundant data. Therefore, it is necessary to perform fossil collection on the logged events. The fossil collection should (1) ensure that events that any subscriber might miss in case of failure are always available, as well as (2) dispose of events that have been received by all subscribers as soon as possible. The RTI ensures that a federate receives all events with timestamp less than its granted time. Therefore, senders do not need to keep events with timestamp less than the granted time of a receiver.

---

[4]When dealing with interactions, the middleware simply codes the federate ID in the tag of an interaction and decodes the ID on reception.

[5]It is possible that the timestamps of some events to be resent are less than the sender's granted time plus lookahead. In this case, the middleware can be designed to encode the content and timestamps of these events in a special RO message, which can then be decoded in the form of TSO events at the receiver's end.

Based on this fundamental assumption, the main task of fossil collection is to determine which logged TSO events are safe for a sender to dispose of according to its current granted time. A time-constrained federate has an associated Lower Bound Time Stamp (LBTS), which is the timestamp of the earliest possible TSO event that may be generated by any other regulating federate [72]. In the scenario depicted in Figure 4.5, we write the lookahead of the ith federate (*Fed[i]*) as La_i, its current granted time as T_i, and the time of the next request this federate may make to the RTI to advance time as T_i'. We define a maximum "*timestep*" by which *Fed[i]* advances its time in each loop as $\delta_i = T_i' - T_i$. Considering the simplest scenario consisting of only two federates, from *Fed[1]*'s perspective, failure may occur in *Fed[2]* either (1) after *Fed[2]* has been granted time but before *Fed[2]* makes another request to advance time, or (2) after *Fed[2]* has made a request to advance time to $T_2 + \delta_i$ but before the request is granted.

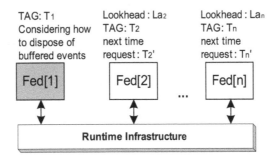

FIGURE 4.5
Example for calculating timestamp of events to be disposed.

Fed[1]'s LBTS is $T_2 + La_2$ in the first case (hence $T_1 \leqslant T_2 + La_2$), and its LBTS is $T_2 + La_2 + \delta_2$ in the second case (hence $T_1 \leqslant T_2 + La_2 + \delta_2$). It is safe for *Fed[1]* to dispose of buffered events earlier than *Fed[2]*'s current granted time T_2, which means any event with timestamp less than $T_1 - (La_2 + \delta_2)$ can be removed immediately.

Generalizing to n federates, suppose that *Fed[k]*($k \neq 1$) has the smallest federate time T_k of the other federates, so that it is safe for *Fed[1]* to dispose of events with time earlier than T_k. It is obvious that $La_k + \delta_k \leqslant \max\{(La_i + \delta_i)|i \neq 1\}$. Thus, in the worst case, it is safe for *Fed[1]* to dispose of all logged with timestamp less than $T_1 - \max\{(La_i + \delta_i)|i \neq 1\}$, and we define this value as *Fed[1]*'s *safe lower bound*. The fossil collection algorithm can easily determine this safe lower bound given that the lookahead and of other federates are available, and this can be achieved easily using a middleware approach.

Furthermore, for any federate, its "timestep" may change from time to time. To minimize global propagations, a simulation time window can be defined with an upper and lower bound specified. The window of a federate (say *Fed[i]*) is an interval around $T_i + \delta_i$, i.e. $[(T_i + \delta_i) - \zeta_1, (T_i + \delta_i) + \zeta_2]$, for some $\zeta_1 < \delta_i$ and $\zeta_1, \zeta_2 > 0$. When *Fed[i]* requests to advance its time to T_i', there are four cases:

1. If $T_i' \leqslant (T_i + \delta_i) - \zeta_1$, set new $\delta_i = (\delta_i - \zeta_1)$.
2. If $(T_i + \delta_i) - \zeta_1 < T_i' \leqslant (T_i + \delta_i)$, δ_i is unchanged.
3. If $(T_i + \delta_i) < T_i' \leqslant (T_i + \delta_i) + \zeta_2$, set new $\delta_i = (\delta_i + \zeta_2)$.
4. If $T_i' > (T_i + \delta_i) + \zeta_2$, set new $\delta_i = (T_i' - T_i)$.

When δ_i decreases, it is safe for the other federates to calculate safe lower bounds using a larger δ value for *Fed[i]*. In case (1), we still need to send other federates the new δ_i, as it is out of the window. For cases (3) and (4), using middleware can ensure that other federates have received the new δ_i before *Fed[i]* requests the RTI to advance time. After *Fed[i]* is granted a new time, the time window will be moved forward to adapt to the change.

4.3.4　Optimizing the Failure Recovery Procedure

The failure recovery procedure starts from the point where a failure is detected by the first federate and ends at the point where all federates are completely re-initialized and ready for resuming normal execution. The straightforward recovery scheme (see Figure 4.3) requires two time-consuming RTI related operations to be performed: (1) to create the physical federation and (2) for each federate, to join the existing federation. The *joinFederationExecution* call incurs costly federation-wide operations. For example in DMSO RTI-NG, this operation usually requires opening TCP sockets to all other federates in the federation, which is expensive [95]. To minimize the overhead (which can be greater than 20 seconds; see Section 4.4), a possible solution is to avoid these calls during the procedure itself. We attempt to solve this problem using a *Physical Federate Pool* approach as shown in Figure 4.6.

This approach creates one or multiple "backup" physical federations[6] concurrently with the normal simulation execution. An appropriate number of PhyFed instances are created, which join their respective backup federations and form PhyFed instance pools (one pool for each federation). In the context of the pool approach, a PhyFed instance may operate in two modes: (1) *working mode*, servicing a virtual federate as normal; and (2) *idle mode*, calling tick regularly to maintain connection session with the RTI while checking for invocation from a virtual federate. On start-up, a virtual federate connects to a PhyFed from the pool and the PhyFed operates in working mode from

---

[6] Depending on the fault-tolerance requirements, these multiple physical federations can be supported by one or multiple RTIEXEC and the backup physical federates can be executed on the same or different machines as the active physical federate.

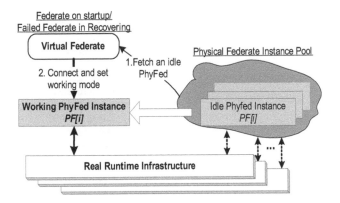

FIGURE 4.6
Physical federate pool approach

then onward. The backup physical federations purely consist of idle PhyFeds instances, which are neither time regulating nor time constrained, and only have minimum interaction with each other. The backup physical federations potentially serve for recovery in the future. On failure recovery, an idle PhyFed instance can be fetched from the pool by the virtual federate to provide the required RTI services immediately. Thus, this approach avoids consuming time for creating the federation and joining the federation execution prior to state replication. Maintaining spare PhyFed instances consumes extra system resources, and we need to investigate the overhead this may cause. Correspondingly, the straightforward fault recovery scheme (Section 7.1) can be optimized using the pool approach as shown in Figure 4.7.

Another uncertain factor is the time needed for the remaining federates to detect the failure propagated from the origin. It depends on the form in which the failure appears and how the fault-tolerant model handles it. If the failure is detected as one of the last three cases defined in Section 4.3, other federates' middleware need only immediately initiate a passive failure recovery. For the first case, the time required to confirm the occurrence of a failure must be longer than the specified "time-out" period.

The situation becomes even more complicated if the "symptoms" of failure cannot be explicitly identified at all. For example, suppose a federate does not receive a *timeAdvanceGranted* (TAG) for a significantly long period after it requests advancing its time from the RTI [1]. Basically this may due to the fact that (1) some LRCs have failed, or (2) the condition for granting its request has not been met yet, or (3) some other reason not related to failure, for example, an unexpected communication delay for the RTI to convey callbacks. There needs to be a method to distinguish the first case from the others. The PhyFed pool approach can be used to solve this issue: a prese-

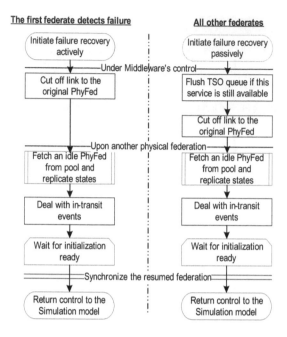

FIGURE 4.7
Illustration of the optimized failure recovery procedure using PhyFed pool.

lected backup physical federation can also serve as an out-of-band channel for a failed federate to notify the remaining federates of the occurrence of failure. We define a special "system" object class ("RTI_FAIL") and have all idle PhyFed instances subscribe to and publish it. The Management Module of the first failed federate registers an RTI_FAIL object instance in the selected backup physical federation. The remaining federates' Management Modules periodically check the existence of such an object from the selected backup physical federation to decide whether to start a passive fault recovery procedure. Hence, it is possible for the whole federation to quickly respond to a local failure.

4.4 Experiments and Results

In order to verify the correctness and investigate the overhead incurred in the proposed fault-tolerant model, we perform a series of experiments to compare the robust federates with normal federates using a simple distributed supply chain simulation.

4.4.1 Configuration of Experiments

The simulated supply chain comprises an agent company, a factory, and a transportation company. The agent keeps issuing orders to the factory, and the latter processes these orders and plans production accordingly. The transportation company is responsible for delivering products of the factory and reporting the delivery status. The three nodes in the supply chain can be modeled as three federates as shown in Figure 4.8, namely *simAgent, simFactory,* and *simTransportation.* These federates form a simple distributed simulation to simulate the supply chain's operation in almost a year (from simulation time 0 to 361). Two object classes "Order" "Products" and one interaction class "deliveryReport" are defined in the Federation Object Model (FOM) [65] to represent the types of events exchanged among the federates. Table 4.1 gives the classes published and/or subscribed to by the federates. The *simFactory* reports the cost incurred for each order at the end of the simulation. The simulation starts with an initialization procedure and then enters the "real" simulation procedure after a global synchronization. The initialization procedure denotes the interval from the point a federate is started to the exact point where it has completed the following operations: create/join the federation, enable time-regulating/constrained, publish/subscribe object/interaction classes and register object instances. During the simulation procedure, federates interact and coordinate time advancement with each other using the conservative synchronization scheme. In this chapter, the elapsed times of the initialization procedure and the simulation procedure of each run are called its initialization time and simulation execution time, respectively.

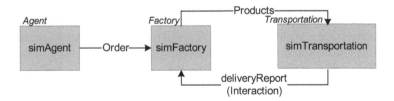

FIGURE 4.8
A simple distributed supply-chain simulation.

Using the same codes for the simulation models, the federates are built into two versions by linking to (1) the DMSO RTI library directly (normal) and (2) the RTI++ middleware library supporting fault tolerance (robust). The RTI++ in these experiments adopts the PhyFed pool approach and uses the IPC Message Queue [101] as the external communication to bridge the virtual federate and its PhyFed. The PhyFed pool maintains one backup physical federation consisting of three idle PhyFeds.

TABLE 4.1

Declaration Information of the Federates.

Federate	Object Classes and Attributes		Interaction Classes and Parameters
	Order	Products	deliveryReport
	Index,Size	*Amount,Index,Date*	*Index,Status*
simAgent	Publish	NIL	NIL
simFactory	Subscribe	Publish	Subscribe
simTransportation	NIL	Subscribe	Publish

The experiment architecture and platform specification are listed in Table 4.4.1. The experiments use three up to twelve workstations and one server, which are interlinked via a 100Mbps-based backbone. Workstations 4 through 12 are only used in scalability studies (Section 3.4). Each federate occupies one individual workstation, with the RTIEXEC and FEDEXEC processes running on the server.

4.4.2   Correctness of Fault-Tolerant Model

To verify the correctness of the fault-tolerant model, we specify a federate *sim-Agent* to generate the same set of orders in different runs. There are three sets of experiments in this session. We first execute the normal federates, in which the outputs are used as a reference in subsequent experiments. Second, we repeat the simulation using the robust federates without introducing failure (FAULT_FREE). The last experiment also uses robust federates but with failure abruptly triggered once by manually terminating a working PhyFed during the simulation procedure (FAULT_INCURRED). The outputs obtained using normal federates are summarized as follows:

1. *simAgent* issues 240 orders, in which the first and the last order carry timestamp 2.5 and 362.5, respectively.

2. *simFactory* receives 239 orders (note that the last order is not received as it is after the simulation end time) and makes products accordingly.

3. *simTransportation* receives all product updates issued earlier than the end time and sends deliveryReport interactions with respect to these updates.

From the FAULT_FREE and FAULT_INCURRED experiments, we check the orders issued and received, products produced and delivered, as well as

TABLE 4.2

Configuration of Experiment Test Bed

Specification	Computers			
	Workstation 1~2	Workstation 3	Server	Workstation 4~12
Operating System	Sun Solaris OS 5.8	Sun Solaris OS 5.8	Sun Solaris OS 5.8	Sun Solaris OS 5.9
CPU	Sparcv9 CPU,at 900 MHz	Sparcv9 CPU,at 360 MHz*2	Sparcv9 CPU*6 at 248 MHz	Sparc II CPU,at 400MHz
RAM	1024M	512M	2048M	512M
Compiler	GCC 2.95.3	GCC 2.95.3	GCC 2.95.3	GCC 2.95.3
Underlying RTI	DMSO NG 1.3 V6	DMSO NG 1.3 V6	DMSO NG 1.3 V6	DMSO NG 1.3 V6
Processes running on	or SimTrans-portation	simFactory	RTIEXEC & FEDEX	SimAgent or SimTrans-portation

the calculation of costs. Outputs (including the timestamps and values of all events) in these experiments match exactly those using normal federates. This indicates that the fault-tolerant model does not introduce any variation to the simulation results, and our framework provides a correct robustness mechanism for HLA-based distributed simulations. The FAULT_INCURRED experiments also show the benefit of the decouple architecture. The failure and the recovery procedure are properly handled and executed by middleware during the runtime. The fault handling and recovery is transparent to the simulation model execution. The user's simulation model was executed exactly in the same way as if the failure had not occurred.

4.4.3   Efficiency of Fault-Tolerant Model

To investigate the performance of the fault-tolerance mechanism, another set of experiments are performed to collect the overall execution time using normal and robust federates. We specify federate *simAgent* to generate orders randomly in each run. For normal federates, we have a number of runs, and the average execution time of these runs is referred to as the NORMAL time of executing one simulation session. As for the robust federates, we first repeat the FAULT_FREE experiments and then carry out a number of FAULT_INCURRED experiments. From FAULT_INCURRED experiments, we select three runs in which the failure of the PhyFed corresponding to federate *simFactory* occurs only once at simulation time 43, 182, or 320. These points represent failure at the start (FI_S), middle (FI_M), and end (FI_E) stages, respectively.

The average CPU utilization of a single normal federate or a virtual federate (in Workstation 1 or 2) is reported as above 80%. A PhyFed has an average CPU utilization as low as <0.5% in working mode and <0.02% in idle mode.

The initialization time of normal federates varies between 19 and 27 seconds in different runs, and it varies between 21 and 27 seconds using robust federates. The latency for initiating the PhyFed pool is well hidden. The simulation execution times of different experiments are reported in Figure 4.9. The normal simulation execution time is ~584 seconds using normal federates, which is almost the same as the average simulation execution time in FAULT_FREE experiments. This means the overhead for federate decoupling and maintaining the PhyFed pool has little influence on execution efficiency. In the FAULT_INCURRED experiments, the simulation execution time is only 11 to 13 seconds longer than the normal case. The overhead does not fluctuate much for failure at different stages of the simulation execution. In the experiments, if an RTI call issued from the virtual federate has not been returned by the PhyFed after more than 6 seconds, a time-out will occur. Because of this, a large part of this slight overhead is due mainly to the failure detection procedure.

When failure occurs, we assume that the normal federates have to start

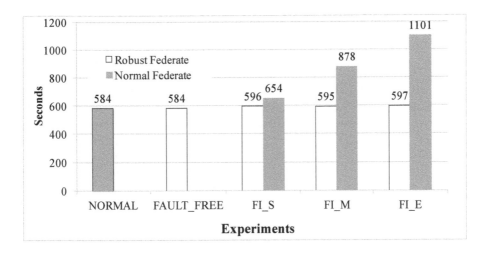

FIGURE 4.9
Simulation execution time of different experiments.

from the beginning, and the sum of the elapsed times of both the failed and repeated simulation executions is used for comparison with the simulation execution times using robust federates. The percentage of saved execution time is shown in Figure 4.10. Obviously, the later the failure occurs, the more execution time can be saved (up to 50%).

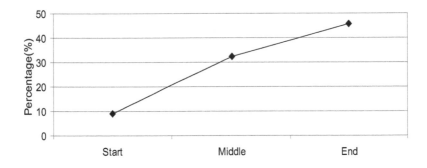

FIGURE 4.10
Percentage of saved execution time with failure occurring at different stages.

4.4.4 Scalability of the Fault-Tolerant Model

A third series of experiments is performed to test the scalability of the fault-tolerant model. This consists of ten sets of experiments in total, with the number of federates varying from three to twelve. Each set of experiments uses a federation that always contains one *simFactory* and one or multiple instances of *simAgent* and *simTransportation*. In the first set of experiments, the federation has one *simFactory*, one *simAgent*, and one *simTransportation*. In each subsequent set of experiments, we always introduce one new instance of *simAgent* or *simTransportation* alternately to the previous set of experiments. As shown in Figure 4.11, each added *simAgent (simTransportation)* is marked with the total number of federates in the federation after it is added. Similar to the experiments reported in Section 3.3, the execution time is measured and compared using different types of federates, that is, normal and robust (FAULT_FREE and FAULT_INCURRED) federates. For FAULT_INCURRED experiments, the failure of federate *simFactory* is configured to occur only once at simulation time 182 (i.e., always at the middle of the simulation execution). The experiment architecture and platform specification for the scalability study are listed in Table 4.1, and the workstation on which each federate operates is also illustrated in Figure 4.11.

The average CPU utilization of each federate is the same as in previous experiments. The execution times for all experiments are recorded in Figure 4.12. The execution times using normal federates increases smoothly when the federation contains an increasing number of federates (starting from 584 seconds for three federates to 714 seconds for twelve federates), and the same trend can be observed when using robust federates. The extra execution time (versus normal federates) consumed by FAULT_INCURRED federates are highlighted in Figure 4.13. Note that the percentage increment of the execution time remains invariant to the number of federates in the federation. It is about 2% for all the cases. The experimental results show that (1) the fault-tolerant model scales well with increasing distributed simulation size; (2) in the case of no fault, robust federates perform almost the same as the normal federates; and (3) in the case of a fault occurring, the model's overhead remains negligible compared to normal federates encountering no fault.

Figure 4.14 shows the overall simulation execution times of normal federates and robust federates in the present of the fault. Similar to the experiments reported in the previous subsection, we assume that the normal federates must restart from the beginning when the fault occurs. Using the same calculation method as in Figure 4.10, the percentages of saved execution time are given in Figure 4.15. The percentage remains about 32% steadily with increasing federation scale. The results indicate that the robust federates can significantly reduce execution time when dealing with failure, compared to normal federates without fault-tolerance functionalities.

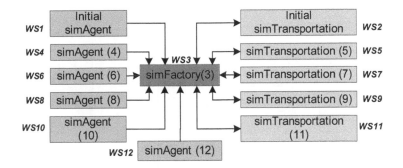

FIGURE 4.11
Initial federation for the ten sets of experiments for the scalability test.

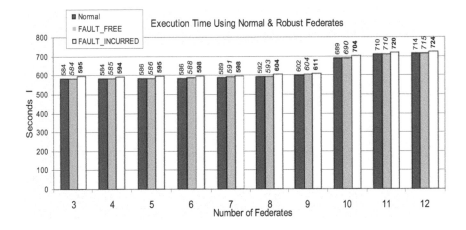

FIGURE 4.12
Simulation execution times with increasing number of federates.

4.4.5 User Transparency and Related Issues

As introduced previously, the extra effort needed for building robust federates is minimal as users only need to link normal federates' code to the RTI++ library. Our framework does not require a federate to be modeled on any particular software package or specially coded for the purpose of fault tolerance. The FOM defined for a normal federate only needs to be slightly extended for the use of the corresponding robust federate, which includes several extra "system" object/interaction classes. Table 4.4.5 lists the system object/interaction classes added into the FOM.

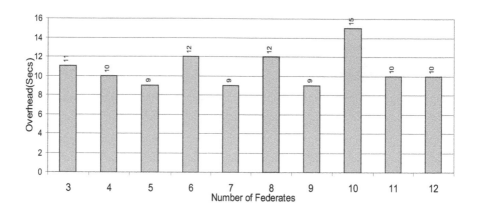

FIGURE 4.13

Overhead of robust federates (fault incurred) versus normal federates without faults.

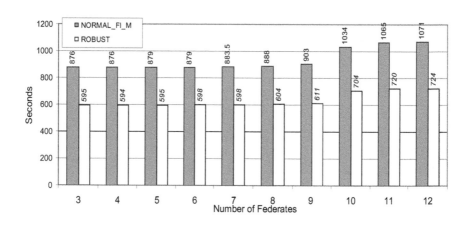

FIGURE 4.14

Simulation execution times with failure incurred in the middle stage.

To examine whether robust federates can interoperate properly with normal federates, we repeat the same simulation scenario as described in Section 3.2 using both robust federates and normal federates in each session (twenty-three possible combinations in total, inclusive of two constructed by pure

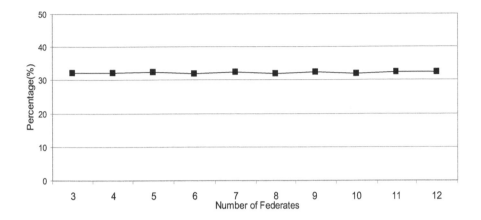

FIGURE 4.15
Percentage of saved execution time with increasing number of federates (fault occurs at middle stage).

normal/robust federates). The extended FOM applies for both types of federates. The six "hybrid" sessions have the same outputs as those reported in Section 3.2. Robust federates upon our framework can interact properly with normal federates. The extended FOM does not cause any semantic problem or any difference in simulation execution. The experiments further show that the fault-tolerant model provides reuse of federate code and user transparency.

4.5 Summary

In this chapter, we introduced a framework for supporting runtime robustness to HLA-based distributed simulations. We investigated the issues and design of a generic fault-tolerant federate model. Based upon the Decoupled federate Architecture, the model was developed to prevent an RTI error from disrupting the execution of simulation federates and ensuring correct recovery of a distributed simulation session. Algorithms were presented to ensure reliable delivery of in-transit messages as well as perform safe fossil collection.

The fault-tolerant model supports the reuse of legacy federates while enabling robustness, and it minimizes developers' efforts for modeling robust federates. The model is platform neutral and model independent. User trans-

TABLE 4.3

System Object/Interaction Classes in the Extended FOM

Object/Interaction Classes	Declaration	Functionalities
RTI_FAIL(Object)	Publish&Subscribe	Notifying other federates about the occurrence of an RTI failure
SYS_FED_ DECLARATION (Interaction)	Publish&Subscribe	Broadcasting the information about which object/interation classesthe local federate has published/subscribed to
SYS_IN_TRAN_MSG (Interaction)	Publish&Subscribe	Delivering the content of intransit TSO or RO events to the receiver during fault recovery

parency has been provided with failure properly masked. Robust federates do not require rollback of simulation execution in the case of failure.

A series of experiments was performed to investigate the correctness and performance of the fault-tolerant model using an example of a distributed supply chain simulation. The experimental results were compared for normal and robust federates in terms of uniformity of output statistics and computing efficiency. The output statistics indicate that the model provides correct fault recovery. The results show that robust federates have a very close performance to normal federates and only incur minimal extra overhead. Our work indicates that the fault-tolerant model is a feasible and efficient solution to the support of runtime robustness in HLA-based distributed simulations, which can be used in the design of robust RTI software in the future.

5

Synchronization in Federation Community Networks

CONTENTS

A large-scale High Level Architecture (HLA)-based simulation can be constructed using a network of simulation federations to form a "federation community." This effort is often for the sake of enhancing scalability, interoperability, and Composability, and enabling information security. Synchronization mechanisms are essential to coordinate the execution of federates and event transmissions across the boundaries of interlinked federations. We have developed a generic synchronization mechanism for federation community networks with its correctness mathematically proved. The synchronization mechanism suits various types of federation community networks and supports reusability of legacy federates. It is platform neutral and independent of federate modeling approaches. The synchronization mechanism has been evaluated in the context of the Grid-enabled federation community approach, which allows simulation users to benefit from both the Grid computing technologies and federation community approach. A series of experiments has been carried out to validate and benchmark the synchronization mechanism. The experimental results indicate that the proposed mechanism provides correct time management services to federation communities. The results also show that the mechanism exhibits encouraging performance in terms of synchronization efficiency and scalability.

5.1   Introduction

Distributed simulation technology facilitates the construction of a large-scale simulation with simulation components of various types, which can be developed independently and distributed geographically. The High Level Architecture (HLA) defines the rules, interface specification, and object model template to support reusability and interoperability among the simulation components, known as federates [33]. While the HLA serves as the de facto standard for distributed simulations, the Runtime Infrastructure (RTI) software provides services[1]to support and synchronize the interactions among different federates conforming to the HLA standard to sustain an overall simulation application, known as a federation as shown in Figure 5.1(A).

In the case where the problem domain is particularly complex or involves multiple collaborative parties, the analysts often need to construct large-scale HLA-based simulations that may involve a large number of federates and vast computing resources over a network or the Internet. Some typical examples are military commission rehearsal, Internet gaming, biology simulation, and supply-chain simulation. Sometimes such large-scale simulations need to be constructed upon multiple simulation federations. Despite the tremendous advantages brought by the HLA technologies, the HLA standard does not explicitly sustain interoperation between federates across the boundaries of federations. To address this issue, a method has been proposed to harness a network of federations to achieve a common goal in the form of a "federation community" [88]. Figure 5.1(B) illustrates an example federation community network. In addition to the advantages of using flat federations[2], simulation developers and users can benefit from the federation community method in (1) improving the scalability of large and complex applications by reducing the

---

[1]A total of six service categories are defined in the IEEE 1516 HLA standard, namely Federation Management, Declaration Management, Object Management, Ownership Management, Data Distribution Management, and Time Management.

[2]To distinguish a traditional federation from federation communities, a simulation upon an individual federation is called a flat federation.

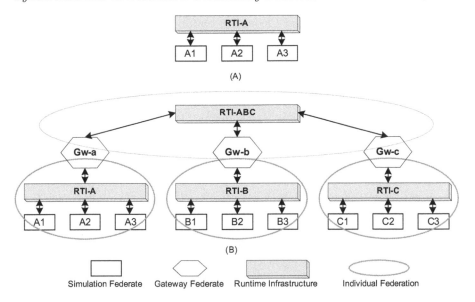

FIGURE 5.1
Illustrating (A) flat federation and (B) federation community.

bandwidth requirements through the partitioning of network load and filtering out irrelevant data among federations, (2) enhancing the Composability of simulation by enabling simulation development using legacy federations, (3) supporting interoperation between heterogeneous federations and RTIs, and (4) facilitating information security [9, 10, 11, 14, 78, 88]. The applications of federation community have been extensively discussed in existing literature, including the architecture for information hiding adopted in secured supply-chain simulation developed by our team [11].

The emergence of Grid computing technologies meets the requirement of large amounts of distributed computational and data resources by the increasing size and complexity of simulation applications. The Grid provides a flexible, secure, and coordinated resource that sharing environment that can facilitate distributed simulation execution [95]. In the past few years, there has been increasing interest in taking advantage of the Grid technologies to execute HLA simulations over the Internet. A few research groups have successfully enabled the Grid technologies to RTI, which either use middleware approaches to encapsulate vendor-provided RTIs [16, 95, 113, 118, 122] or implement the RTI directly using Grid services [40, 91]. While the existing Grid-enabled simulation techniques have proved highly advantageous, users can benefit even more if a whole federation is accessible through Grid services. Once a Grid-enabled federation community is constructed, the computational resources within administrative domains can be further exploited with the reusability[3] of legacy federates maximized when enabling the Grid technology.

Simulation federates in a federation community should be provided with RTI services as in a flat federation [18] so that they can interact with each other and the RTIs involved, irrespective of whether or not those federates operate in the same federation. Support to the RTI services related to Federation Management, Declaration Management, Object Management (OM), and Data Distribution Management (DDM) has been well addressed to facilitate the execution of federation communities [9, 10, 76, 77]. Nevertheless, a more important and challenging issue, namely Time Management (TM) over federation community networks, has received little attention. Time management is concerned with the mechanisms for controlling the advancement of each federate along the federation time axis and synchronizing event (the terms "event" and "message" are used interchangeably in the rest of the chapter) delivery among federates. Without a properly designed synchronization mechanism, the overall simulation execution upon a federation community is susceptible to state inconsistencies, the federates in different federations receiving messages for the same set of events in different orders. However, the current IEEE 1516 HLA standard is not intended to foster time management across the boundaries of federations [65]. There exists only a few preliminary or nonstandard methods to address this issue. Hence, there is a pressing need for a generic synchronization mechanism for HLA-based federation communities.

This chapter proposes a generic approach to synchronizing federates and events within federation community networks. The proposed synchronization algorithms have been developed based on the gateway federate approach for constructing federation communities. The algorithms operate inside the gateway federates, and the time constraints from any federation are propagated through the whole network to coordinate the progress of the whole federation community along the simulation time axis. The approach has also been successfully applied to Grid-enabled federation communities. The remainder of this chapter is organized as follows: Section 5.2 gives an overview of federation communities and related issues. Section 5.3 discusses existing work and analyzes the challenges to be addressed. Section 5.4 details the algorithms for synchronizing Timestamp Order events crossing federation boundaries and proves the algorithms' correctness. Section 5.5 presents the benchmarking experiments, which examine the correctness of the synchronization mechanism and evaluate its performance. In Section 5.6, we conclude with a summary and proposals for future work.

---

[3]Reusability may be supported at the executive level and/or code level. Reusability at the executive level refers to the feasibility of reusing federate executives directly, while reusability at code level means reusing the legacy federates' source code, such as via middleware approaches.

5.2 HLA Federation Communities

The technology for constructing federation communities has been well studied. Federation communities may be constructed for different objectives (see Section 5.1) and various architectures architectures. The Grid computing technologies can also be combined with federation community approaches to further benefit simulation users.

5.2.1 Construction Approaches

Federation community networks are formed usually via one of the following approaches as defined in [88, 91], that is, gateway federate, federation gateway process, federation broker, or federation protocol. A gateway federate (also referred to as a proxy federate [88] or bridge federate [8, 73], see Figure 5.2(A)) is a specially designed federate that joins multiple federations simultaneously and performs the translating functions between them so as to accomplish the multi-federation interoperability. A federation gateway process (Figure 5.2(B)) is a nonfederate process that interconnects two or more federates in multiple discrete federations and does the translation between these federations. A federation broker (Figure 5.2(C)) is a process that directly connects heterogeneous RTI implementations together through an RTI-to-RTI API to be defined. It passes RTI internal states between multiple RTIs. Instead of

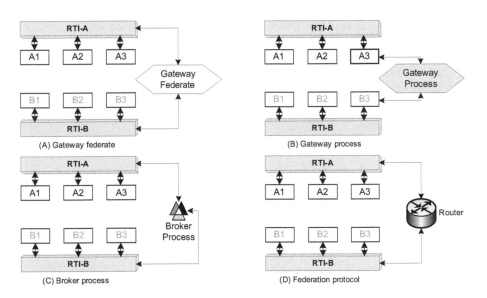

FIGURE 5.2
Typical approaches to construction of federation community.

using any intermediate process as in the above three approaches, the "federation protocol" refers to a lower-level RTI-to-RTI protocol to enable direct communication between heterogeneous RTI implementations (Figure 5.2(D)). Another interesting work is the SIP-RTI proposed by Van Ham [110], which aims to interoperate heterogeneous RTIs via a middleware layer below the RTI APIs and successfully enables Ownership Management services.

Generally speaking, the functionalities of a federation gateway process (Figure 5.2(B)) can be fully supported by a customized gateway federate (Figure 5.2(A)). Although the gateway federate approach cannot access low-level RTI information as the broker or protocol approach does, it does not intrude on the execution of simulation federates or demand any extra support from the RTI software in addition to the standard RTI services. The approach also frees modelers from revising legacy federate code or having to use designated middleware. The performance of the approach has proved acceptable in terms of latency [10].

A gateway federate is normally available as a single process having multiple Local RTI Components (LRC) [1]. In some cases, we can also couple multiple processes (possibly distributed) together to form a gateway federate executive in a general sense, such as using the Decoupled federate Architecture to incorporate the Grid computing technologies [21], for example, the gateway federates used for Grid-enabled federation communities (see Section 5.4).

Capitalizing on the merits of the gateway approach, the work presented in this chapter utilizes gateway federates for constructing federation communities. In particular, our design assumes an exclusive gateway for a user federation (composed of users' simulation federates and a gateway federate) while interlinking multiple gateway federates via additional gateway federations. Clearly, a gateway federation consists of only gateway federates instead of any simulation federate. From any simulation federate's point of view, the other federates inside the same federation are its local federates, while those in other federations are its remote federates. The federations that a gateway federate participates in are termed neighbor federations to each other.

5.2.2   Architectures of Federation Community Networks

The architectures of federation communities are indefinite, depending on the users' requirements. Some typical examples are peer-to-peer federations (Figure 5.3(A)), set-based federations (Figure 5.3(B)), and hierarchical federations (Figure 5.3(C)). Figure 5.3(D) illustrates another example with a more complicated architecture. For the first three cases, there exists only one unique path for message delivery from a federate to any remote federate. In a federation community with structure as shown in Figure 5.3(D), two remote federates can be connected via multiple paths, for example, path 1 and 2 between federate C2 and E2; thus, loops can be formed in this federation community. Inside a federation, any message generated by a federate is reachable to any other federate through the RTI, and it is not necessary to specify any receiver. This feature makes a federation similar to a subnet with the federates and the RTI resembling the nodes and the hub, respectively [105]. Therefore, a federation community network can always be viewed as interconnected Local Area Networks (LANs). In this sense, a gateway federate resembles a bridge or switch supporting local internetworking, which performs the functionality of routing messages among federations. Therefore, to address the problems caused by loops in a network, the spanning tree algorithm used for interconnected LANs has been effectively transplanted to construct a loop-free topology for a federation community network [9]. Hence, our approach assumes the support of gateway federates for loop-free messaging in any federation community network. Undoubtedly, the leaf nodes will always be user federations.

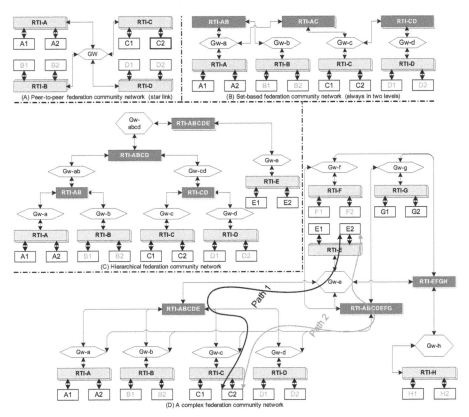

FIGURE 5.3
Illustrating various architectures of federation community networks.

A gateway federate may transfer messages from/to any pair of neighbor federations, such as the star link shown in Figure 5.3(A). There does not exist any technical barrier to using a gateway for multiple-user federations either. Nevertheless, to simplify the design of control mechanisms, this study assumes that a user federation only has one gateway federate. Neighbor federations will be layered according to their distances to the leaf nodes, the user federation, or the nearest gateway federation will be at the lower layer and the remainder will be at the upper layer. More details are available in Section 5.2.2.1 and 5.4.1. Messages will be forwarded only between the lower federation and any of the upper federations[4].

5.2.2.1   Proposed Internal Architecture of the Gateway Federates

In [9, 10, 76, 77], we have detailed the design and implementation of the gateway federate in connection with Federation Management, Declaration Management, Object Management, and Data Distribution Management. This chapter is only concerned with the two basic functionalities directly servicing the need of Time Management, that is, routing and synchronization (Figure 5.4). Routing deals with delivering a message to the right federations (then the recipients), and synchronization is about delivering these messages in the correct order without violating the HLA rules and coordinating time advances of all federates in the community. By making use of the declaration management and object management services, an accurate routing algorithm (detailed in [13]) has been developed to map the publishing, object instance registration, and subscription information of federates residing in different federations.

Figure 5.4 depicts a portion of a gateway federate's internal architecture using the gateway federate GW-E in Figure 5.3(D) as an example. The gateway federate comprises multiple "federate modules," a TSO event buffer, a routing module, and a synchronization module. Each federate module operates an individual LRC to directly interact with the respective federation.

Assuming the gateway federate is dealing with TSO events originating from RTI-E, the routing module maintains the events in the buffer and refers to the routing table to direct each event to the neighbor federation(s) when needed. Thus, these events may be gradually relayed to the remote destination federations. The synchronization module then decides at any point which events should be relayed according to their timestamps. The synchronization module also propels the federate modules to advance times in all neighbor federations. The basic idea is to impose the time constraints from each federation to the neighbors, and other gateways propagate the constraints through the whole network. The whole federation community then progresses along the simulation time axis in a coordinated manner. The synchronization algorithms are detailed in Section 5.4.

5.2.3   Grid-Enabled Federation Community

While the HLA enables the construction of large-scale simulations using distributed simulation components, the Grid technologies enable collaboration

---

[4]This does not apply in star link federation communities, as all federations are peer-to-peer linked by a single gateway federate.

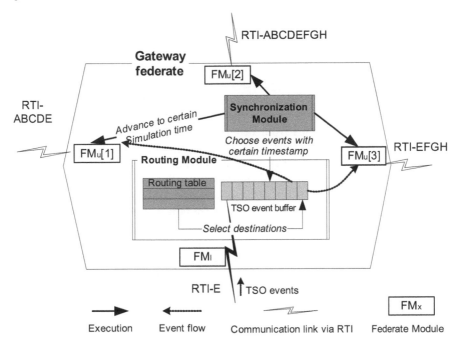

FIGURE 5.4
A simplified view of the internal architecture of a gateway federate.

and provide mechanisms for the management of distributed computing resources where the simulation is being executed, while also facilitating access to geographically distributed data sets. The Grid computing technologies have been employed to present HLA-compliant simulation models as Grid services to the external simulation users outside an administrative domain.

Although these existing Grid-enabled distributed simulation systems have exhibited significant merits, there are still two issues to be addressed due to the existing security rules of most administrative domains. One issue is that only a few nodes of an administrative domain are accessible to the external users. As an indicative example, the HLA_Grid_RePast framework [16] supports large-scale agent-based simulations and has been deployed on top of two high-performance clusters, one at the Parallel and Distributed Computing Center (PDCC) in Singapore and another at the Midlands e-Science Center in Birmingham (UK). In this test bed, only the simulation model running on the master node of each cluster can be accessed by external users, while the models hosted by a number of other worker nodes are isolated by the firewall. However, we can envision some customized approaches using distributed computing technologies, such as MPI, to make use of the computing resources on the internal nodes. Those will either be application specific or require considerable reengineering of the existing simulation federates.

Another pending issue is that federate developers often already have a set of federates situated over their intranet; thus, it is desirable for users to access a whole federation through Grid services to further exploit computational resources and support the reusability of federates at the executive level[5]. In an HLA-based simulation, federates interact only through the RTI. Federates are not aware that other federates are connected in the same federation. From a federate's point of view, its correct execution can be guaranteed as long as the events it consumes from other federates and the events it generates to them have correct content and order. This conforms to the fundamental principle of event-driven distributed simulation. Thus, we can use a gateway federate to collect the events of the federates from the intranet and present the gateway federate as a set of Grid services on a public node. Eventually, the resources inside the administrative domain and the internal simulation federates are exposed to the authorized external users.

Two simple examples of Grid-enabled federation community are shown in Figure 5.5, in which each gateway consists of one federate module interacting with local RTI and another Grid-enabled federate module driving a proxy running in a federation controlled by the user. The Grid-enabled module calls the RTI services through a standard RTI interface, and the calls are translated to Grid invocations over the Grid computing backbone to drive a proxy process [16]. A proxy is a special federate process that contains the real LRC and executes whatever RTI services instructed by the Grid-enabled module. The gateway then interacts with Grid-enabled federations as if those are supported by normal RTIs.

5.3   Time Management in Federation Communities

In the context of HLA-based simulations, a federate can be both time regulating and time constrained, either of them, or neither. Time-regulating federates may send timestamp ordered (TSO) events while time-constrained federates are able to receive TSO events in time order. Each regulating federate must specify a "lookahead" value and ensure that it will not generate any TSO event earlier than its current time plus lookahead [65]. After requesting a time advance from the RTI and getting granted, the federate is passed to all events with timestamp less than or equal to the federate's granted time.

The HLA standard defines two synchronization approaches, namely a Conservative approach and an optimistic approach. The Conservative approach implements the TSO event delivery services and is used to advance logical time. The principal task of this approach is to determine the lower bound

---

[5]In contrast, the reusability support at code level demands the availability of federate source code.

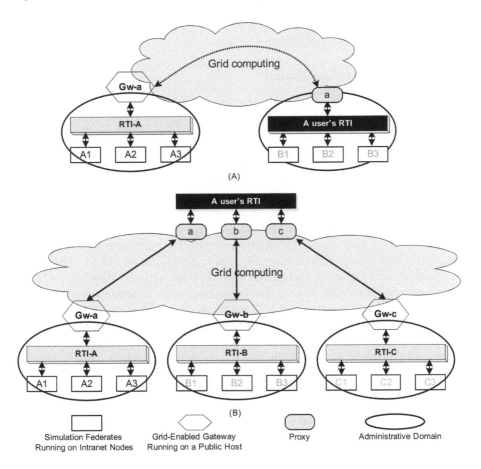

FIGURE 5.5
Two examples of Grid-enabled federation community.

on the timestamp of future TSO events that it will receive [42]. A time-constrained federate has an associated LBTS[6], which is the timestamp of the earliest possible TSO event that may be generated by any other regulating federate [1]. When a federate adopts the Conservative approach, the RTI ensures that no message is delivered to the federate "in its past", that is, no TSO event is delivered that contains a timestamp less than the federate's current time. In contrast, the optimistic synchronization approach allows message processing out of time stamp order, and optimistic federates may enjoy the advantages of being freed from the constraint of other federates. However, because the RTI no longer guarantees orderly event delivery, this approach

---

[6]IEEE 1516 [43] defines a term "Greatest Available Logical Time (GALT)" as the substitute for LBTS. This chapter still uses LBTS to avoid confusion with existing work.

demands dedicated rollback procedures within the simulation models in the case of receiving "past" events.

5.3.1   Problem Statement

A number of research challenges must be addressed in the development of a generic synchronization mechanism. To develop a correct synchronization mechanism among multiple federations is not trivial, as the HLA does not support crossing-federation synchronization at all. It is even more challenging to develop a generic synchronization mechanism to address the drawbacks of the existing work. The HLA only defines the functionalities that the time management services should possess, while it suggests no design rules for the implementation of these services. It is also difficult to probe the internal execution of an RTI during runtime because most RTI implementations operate in a black-box manner. To develop a suite of Time Management services to replace the existing RTI library requires tremendous efforts in spite of the difficulties of integrating them into other services without knowing the RTI's implementation details. Even if this is feasible, users have to reengineer legacy federates to adapt to the new RTI software or middleware. The strategy is prone to failure when the legacy federates are only available as binary executives. It is desirable to have a synchronization mechanism that can reuse the Time Management services built into flat federations and does not intrude on the internal execution of RTIs.

A gateway federate does not contain any portion of the simulation model and will generate no event on its own initiative. It needs to be time constrained to receive TSO events from a federation and be time regulating to reproduce them in other federations, and vice versa. Some existing works rely purely on LBTS calculation to manage the time advance of correlated federations, which will encounter the problem as follows:

- Considering the simplest federation community consisting of only two federations, each containing only one simulation federate. In the scenario depicted in Figure 5.6, we specify the lookahead (La) of both federates as 1.0, and the "*timestep*" by which every federate advances its time in each iteration as 1.0. When a federate ($Fed[i]$) is granted to time T_i, it generates an event with timestamp $T_i + La_i + 0.5$ (written as $Event_i$ $(T_i + La_i + 0.5)$), and this event should be delivered to another federate. Both federates start with simulation time 0. We mark the federate module interacting with RTI-A as FM_a and the one interacting with RTI-B as FM_b, and zero lookahead is set to both modules.

- Initially, $LBTS_a$ equals $T_1 + La_1 = 1.0$, as does $LBTS_b$. FM_a requests to advance Ta to $LBTS_a$, that is, 1.0, and it will be granted to time 1.0 immediately, as will FM_b be. $Fed[1]$ generates $Event_1(1.5)$ and then will be granted to 1.0 since $T_a = 1.0$ already, same will Fed[2] be. Now $LBTS_a$ and $LBTS_b$ have increased to 2.0, if FM_a requests time advance to 2.0, it will be granted to 2.0 and RTI-A will deliver $Event_1(1.5)$ to FM_a. At this point, if FM_b has not issued the time advance, it still can reproduce $Event_1(1.5)$ through RTI-B. But FM_a will never stand a chance to reproduce $Event_2(1.5)$ through RTI-A as $T_a = 2.0 > 1.5$, which means that $Event_2(1.5)$ is in FM_a's past. To generate an event in a federate's past will inevitably induce state inconsistency. Alternative methods are expected to fix the severe flaws of the existing work.

The ultimate goal of synchronizing federation communities is to ensure that a federate advances its logic time and generates/consumes TSO events in the manner as if all federates are operating in a flat federation. In other words, a well-designed synchronization mechanism is able to properly impose the time constraints of all other federates onto each federate. According to the above analysis, a more reasonable approach should focus on dealing with the TSO events, to coordinate the operations of a gateway in multiple federations based on the status of events.

Furthermore, regardless of the fact that the federation community approach may improve the execution efficiency for some ad hoc applications, the synchronization and communication in a federation community must go through gateways from one RTI to another. These tasks are directly handled by a single RTI in the flat federation. To achieve an acceptable performance requires a "concise" synchronization mechanism to minimize the overhead introduced by the complexity of synchronization control, which is even more cumbersome when processing through multiple intermediate gateways is involved. Therefore, in our view, the research issue about the efficiency of any synchronization mechanism for federate community is how close it may approximate that of a flat federation rather than how much it may outperform, which is unrealistic.

5.4 Synchronization Algorithms for Federation Community Networks

This section proposes a generic synchronization mechanism that takes advantage of the Gateway approach to constructing federation communities. Two novel synchronization algorithms are introduced that are employed by gateway federates, one for layered federations and another for star link federations (see Figure 5.3). The basic idea of the proposed mechanism is to let a gateway throttle the time advance of the federations linked together by the gateway. The two proposed algorithms acquire the time constraint of each federation and gradually impose the constraint over all the federation community networks. Thus, federates belonging to different federations can properly coordinate with each other in time advance as if they are in a single federation.

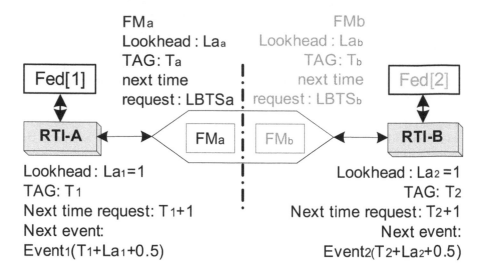

FIGURE 5.6
A simple federation community scenario.

Aiming at generality, the proposed mechanism focuses on the support of Conservative synchronization. This is because Conservative synchronization does not require a Conservative federate's simulation model having the roll-back capability; thus, the Conservative federate is liable to synchronization error when using the optimistic approach. On the contrary, the optimistic federates are capable of interacting with Conservative federates by their nature, and they can be easily revised to use Conservative synchronization anyway. As a matter of fact, the RTI services for optimistic synchronization actually leave the complication to ensure state consistency to the optimistic federate (or middleware) developers. The efficiency of Conservative synchronization is becoming less of an issue with the rapid growth in computing power of computers available to researchers.

Another aim of the proposed mechanism is to provide the designers of future RTI software with a viable direction to address issues related to the construction and execution of federation communities. Our proposed solution does not intend to create a new synchronization approach to substitute the Time Management services of any existing RTI software; instead, it makes the most of the standard services to coordinate federates in multiple federations.

The objectives and scope of this study are summarized as follows:

- To support strictly correct time management services to a federation community regardless of its network structure.

- To provide user transparency, which (1) allows developers to model their federates freely using various software packages and on different platforms; (2) avoids the need for developers to include extra codes in the simulation models to acquire Time Management services for federation communities, thus to minimizing development cost and support reuse of legacy federates at both the code level and executive level; and (3) allows users to deploy/execute federates in the same way as in normal flat federations.

• To minimize overhead for providing synchronization to ensure execution efficiency.

• To facilitate synchronization among simulation models in Grid-enabled federation communities.

5.4.1 Synchronization Algorithms

We first consider those federation communities subject to constraints on message switching (layered). Figure 5.7 presents a fragment of a federation community and highlights the simplified internal view of a gateway federate. The gateway federate (such as Gw-c in Figure 5.3(B)) links one user federation via module FM_l at the lower layer together with multiple gateway federations via module $FM_u[i]$ (i=1, 2, ..., n) at the upper layer (the definition of lower and upper layers available in paragraph 2, Section 5.2.2). The lower layer may be either a user federation or a gateway federation, and the upper layer can only be gateway federations. The TSO buffer maintains the attribute updates and interactions collected from different federations by the federate modules. Let *minNextEventTimeStamp* denote the minimum *timestamp* of all TSO events that may be subsequently delivered to a federate module from the federation it joins, and let *lastGrantedTime* denote the logic time granted to the federate module in the previous iteration, which is also the federate's current time. Particularly, a federate module's *minNextEventTimeStamp* can be obtained via the RTI call *queryMinNextEventTime()* [1]. This value is computed by taking the minimum of the effective federation LBTS and the timestamps of all TSO events (if any) currently queued for delivery to the federate. Four methods, namely *nextEventRequest (NER)*, *nextEventRequestAvailable (NERA)*, *timeAdvancementRequest (TAR)*, and *timeAdvancementRequestAvailable (TARA)*, have been defined in the HLA specification for a federate to request time advance from the RTI. The *nextEventRequest* method advances the federate's logical time to the *timestamp* of the next relevant TSO event in the federation, while the *timeAdvancementRequest* method advances the federate's logical time to a specified timestamp. The other two methods are similar, but they do not guarantee that all events with timestamp equal to the granted time will be delivered to the federate. If a federate with zero lookahead advances its time to T using *timeAdvancementRequestAvailable*, it may advance to time T again in the future without violating the HLA rules, while this is illegal if *timeAdvancementRequest* or *nextEventRequest* has been used.

The features of these methods have been fully utilized in our synchronization algorithms.

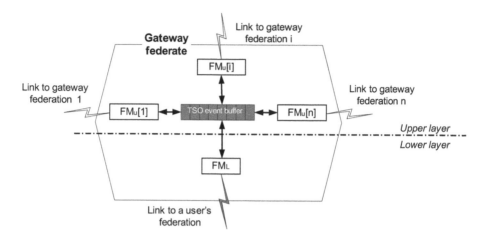

FIGURE 5.7
The fragment of a federation community subject to constraint of message switching.

The synchronization algorithm employed by the gateway federate can be described by the pseudo code in Figure 5.8. We put a section "RTI" in the name of each routine that invokes standard RTI service calls. The algorithm consists of five basic phases:

- **Phase 1**: The federate module FM_l at the lower layer obtains the minimum timestamp of all TSO events that may be subsequently delivered to it from its local federation; the timestamp is referred to as $FM_l.minNextEventTimeStamp$. If FM_l's current federate time (that is, the last time granted) is less than FM_l $.minNextEventTimeStamp$, the control will be passed to the federate modules servicing federations at the upper layer, FM_us. In turn, each FM_u initiates the RTI method *timeAdvancementRequestAvailable* to request time advance to $FM_l.minNextEventTimeStamp$.

- **Phase 2**: In turn, each FM_u keeps invoking the RTI method tick to wait for its local RTI to grant it the requested time advance. On completion of this phase, all FM_us will be granted to a federate time, that is, FM_l $.minNextEventTimeStamp$, and each FM_u enqueues all TSO events received from its local federation (if any) to the TSO event buffer.

- **Phase 3**: The TSO events in the buffer will be processed and routed to the corresponding destination, which is the federation at the lower layer in this case. Control is now with FM_l, which reproduces these events in its local federation.

```
do
//Phase 1: Using the minNextEventTimeStamp of the user federation to advance the logic
time in all gateway federations
      FMₗ.minNextEventTimeStamp= FMₗ.RTI.queryMinNextEventTime()
      if (FMₗ.minNextEventTimeStamp > FMₗ.lastGrantedTime)
         for each module servicing federations at the upper layer, FMᵤ[i], do
              FMᵤ[i].RTI.
         timeAdvancementRequestAvailable(FMₗ.minNextEventTimeStamp)
         end for

// Phase 2: Ticking in all gateway federations one after another until the corresponding
federate modules granted to a new time
         do
            for each FMᵤ[i] do
            FMᵤ[i].RTI.tick()
            end for
            while time advancement request not granted for any FMᵤ[i]
         end if

// Phase 3: Routing each TSO event collected from the gateway federations, with
timestamps not greater than FMₗ.minNextEventTimeStamp and not less than Min(FMᵤ[i].
lastGrantedTime, i= 1, 2, ..., n)
         do
            route_TSO_event()
            FMₗ.forward_TSO_event()
         while TSO event buffer not empty

// Phase 4: Advancing the time of the federate module at lower layer using the minimum
of granted times of the federate modules at the upper layer. Actually, the granted times
are identical and equal to the minNextEventTimeStamp of FMₗ at the beginning of this
iteration. This routine operates until the federate module detects that its
minNextEventTimeStamp is greater than its granted time.
         minTAG = Min(FMᵤ[i].lastGrantedTime, i= 1, 2, ..., n)
         do
            FMₗ.RTI.timeAdvancementRequestAvailable(minTAG)
            while time advancement request not granted for FMₗ do
            FMₗ.RTI.tick()
            end while
         while  FMₗ.RTI.queryMinNextEventTime()  is  not  greater  than
minTAG

// Phase 5: Routing each TSO event generated from the user federation or the gateway
federation at lower layer, with timestamps not greater than Min(FMᵤ[i].lastGrantedTime,
i= 1, 2, ..., n)
         do
            route_TSO_event()
            for any destination federation i at upper layer, do
            FMᵤ[i].forward_TSO_event()
            end for
         while TSO event buffer not empty

while simulation end time has not been reached
```

FIGURE 5.8

Synchronization algorithm for multiple-layer federation community networks.

- **Phase 4**: The minimum of granted times of the FM_us will be computed, that is, $minTAG$. FM_l keeps invoking the RTI method *timeAdvancementRequestAvailable* to request time advance to $minTAG$ and blocks until it detects $FM_l.minNextEventTimeStamp$ becomes greater than $minTAG$. FM_l enqueues all TSO events received from its local federation (if any) to the TSO event buffer;

- **Phase 5**: The TSO events in the buffer will be processed and routed to the corresponding destination, which are the federations at the upper layers. Each FM_u reproduces these events in its local federation if the federation is a destination.

As for the star link federation communities with no constraint on message switching, a more general algorithm will apply. Given a gateway consisting of n federate modules linking to n peer-to-peer user federations, we denote the *i*th federate module as *FM[i]*. The algorithm can be written as the pseudo code in Figure 5.9, which has only slight differences from the first one. This algorithm consists of four basic phases:

- **Phase 1**: The federate module with the least *minNextEventTimeStamp* is identified, and referred to as *FM[k]*. *FM[k]* keeps initiating the RTI method *timeAdvancementRequestAvailable* to request time advance to *FM[k].minNextEventTimeStamp* blocks until *FM[k]* detects the *minNextEventTimeStamp* becomes greater than *FM[k]*'s current federate time. *FM[k]* enqueues all TSO events received from its local federation (if any) to the TSO event buffer;

- **Phase 2**: The TSO events in the buffer will be processed and routed to the corresponding destinations. Each of the other *FM*s (referred to as *FM[i]*, $i \neq k$) reproduces these events in its local federation if the federation is a destination. Each *FM[i]* then initiates the RTI method *timeAdvancementRequestAvailable* to request time advance to *FM[k]*'s last granted time.

- **Phase 3**: Each gains control in turn. *FM[i]* keeps invoking the RTI method tick to wait for its local RTI to grant it the requested time advance. In this phase, each *FM[i]* enqueues all TSO events received from its local federation (if any) to the TSO event buffer.

- **Phase 4**: If *FM[i]*'s time advance request is granted, the TSO events in the buffer will be processed and routed to the corresponding destinations. Each of the other *FM*s (referred to as *FM[m]*, $m \neq i$) reproduces these events in its local federation if the federation is a destination. Phase 3 and Phase 4 will repeat until the requested time advances of all *FM*s (exclusive of *FM[k]*) are granted.

The core point of the two algorithms is to let gateway federates detect the minimum value of the TSO events in all simulation federations and use this

do
// *Phase 1: Getting the federate module FM[k], which has the minimum of all user*
federations' minNextEventTimeStamps, and advance the federate module's time
until the module's minNextEventTimeStamp increases
      for (i=1, i<n; i++) **do**
    FM[i].minNextEventTimeStamp= FM[i].RTI.queryMinNextEventTime()
    end for

    get   *FM[k],*   *FM[k].minNextEventTimeStamp*   =   *Min(FM[i].*
minNextEventTimeStamp, i= 1, 2, ..., n))
    do
     do
        FM[k].RTI.
timeAdvancementRequestAvailable(FM[k].minNextEventTimeStamp)
        while time advancement request not granted for *FM[k]* **do**
       FM[k].RTI.tick()
     end while
       while *FM[k].RTI.queryMinNextEventTime()* is not greater
than *FM[k].lastGrantedTime*

// *Phase 2: Routing each TSO event collected from federation k, with timestamps*
not greater than FM[k].minNextEventTimeStamp and not less than FM[k].
lastGrantedTime
    do
     route_TSO_event()
     for any destination federation $i \neq k$, **do**
     FM[i].forward_TSO_event()
     end for
     FM[i].RTI. timeAdvancementRequestAvailable(FM[k].lastGrantedTime)
    end for

// *Phase 3: Ticking in all other federations one after another until the*
corresponding federate modules granted to a new time
    do
     for each *FM[i]*, $i \neq k$ **do**
     FM[i].RTI.tick()
     end for

// *Phase 4: Routing each TSO event collected from a federation j, with timestamps*
not less than the FM[k]'s logic time
        if *FM[i]* is granted **do**
       route_TSO_event()
       for any destination federation $m \neq i$, **do**
       FM[m].forward_TSO_event()
         end for
       end if
      while time advancement request not granted for any *FM[i]*
     end if

FIGURE 5.9
Synchronization algorithm for peer-to-peer federation community networks.

value as a "safe lower bound" to control the progress of simulation federates
through the gateways. The RTI time advance methods are used to "inform"
the remainder of the simulation of the safe lower bound at any point.

The above algorithms work with the assumption that a simulation federate either has non-zero lookahead or always advances its logic time to a value greater than its current *minNextEventTimeStamp*. Given that a simulation federate (X) has zero lookahead and attempts to advance its time to its *minNextEventTimeStamp*, apparently the federate's current time (granted the most recently) is not greater than $X.minNextEventTimeStamp$. According to the definitions of and *minNextEventTimeStamp*, the *minNextEventTimeStamp* of FM_l in the gateway federate conforms to the inequation: $FM_l.minNextEventTimeStamp \leq X.lastGrantTime+X.lookahead = X.lastGrantTime \leq X.minNextEventTimeStamp$. Meanwhile, the same inequation also applies for the *minNextEventTimeStamp* of federate X; we have $X.minNextEventTimeStamp \leq FM_l.minNextEventTimeStamp$. This is because FM_l also has zero lookahead and always advances to $FM_l.minNextEventTimeStamp$. Thus, federate X and module FM_l will always attempt to advance to a fixed time, and this inevitably causes deadlock.

5.4.2   Proof

The correctness of a synchronization algorithms covers three aspects, that is, (1) the algorithm is logically correct by indicating that all federates generate TSO events and advance time with compliance to HLA rules, (2) the algorithm proves to be deadlock free, and (3) simulation federates are always guaranteed to receive all TSO events within the granted time and in an orderly manner. The proof is given to the first algorithm from the perspective of a gateway federate that connects to a user federation. We assume there are m federate modules at the upper layer.

5.4.2.1   Compliance to HLA Rules

This issue involves federate modules at the upper layer (FM_u) and the module at the lower layer (FM_l). FM_u simulate TSO events collected by FM_l from the user federation. Let $minNET_n$ be the *minNextEventTimestamp* of FM_l on the initialization of the nth iteration of the algorithm. Let $lowerEve_n$ be the set of the events to be collected by FM_l in the current iteration; we have

$$\forall le \in lowerEve_n, \quad le.Timestamp \geq minNET_n \tag{5.1}$$

Let $FMSet_u = FM_u[i]|i=$ 1, 2, ..., m and TAG_n represent the logic time granted to a federate module in the nth iteration. $\forall FM_u[i] \in FMSet_u, FM_u[i].TARA(minNET_n)$; we have

$$FM_u[i].TAG_n = minNET_n \tag{5.2}$$

Let $upperEve_n$ be the set of events collected by FM_us in the current iteration before FM_l initiates time advance.

$$\forall re \in upperEve_n, FM_u[i].TAG_{n-1} \leq le.Timestamp \leq minNET_n \tag{5.3}$$

$FM_l.TARA(min(FM_u[i].TAG_n))$, from equality (5.2), we have

$$\forall le \in lowerEve_n, \qquad le.Timestamp \geq minNET_n \qquad (5.4)$$

$$FM_l.TAG_n = minNET_n \qquad (5.5)$$

$FM_u[i].forward\_TSO\_event(\forall le \in lowerEve_n)$, from (5.4), $FM_u[i]$ should generate TSO events with timestamp equal to $minNET_n$. From (5.2), the current logic times of FM_u are equal to $minNET_n$. The HLA rules permit any time regulating federate, which advance its time using $TARA$, to generate TSO events with timestamp equal to (current time + lookahead). It is assured that (1) the RTI of user federation has guaranteed the order of the TSO events in the federation, which means the $minNextEventTime$ of FM_l will never roll back to a smaller value, (2) user federates have non-zero lookahead or always advance time with a greater time than its current time (see Section 5.4.1). The algorithm makes FM_us advance time only if FM_l detects (the current $minNextEventTime$) > (the last $minNextEventTime$) (5.6) Let $UtimeAdv_k$ be the value of the kth time to which FM_us advance. As FM_us advances their logic times to the $minNextEventTime$ of the user federation, from inequality (5.6), we can have $UTimeAdv_k > UTimeAdv_{k-1}$. From above, we can conclude that the federate modules at the upper layer advance time and generate TSO events conforming to the HLA rules. FM_l generates TSO events copied from FM_u. In the nth time advance iteration of the gateway, let's consider the point where FM_u have time advances granted and FM_l is about to perform $TARA$. The current logic time of FM_l is $FM_l.TAG_{n-1}$.

According to the algorithm, we know that $FM_l.TAG_{n-1}= FM_u$s' granted time in the last itereation. FM_l issues $FM_l.forward\_TSO\_event(\forall re \in upperEve_n)$. From (5.3), we have $re.Timestamp \geqslant FM_l.TAG_{n-1}$. Undoubtedly, this event generation procedure obeys the HLA rules. In order to indicate that FM_l advances time with compliance to the HLA rules, we need to consider two cases. $minNET_n = minNET_{n-1}$: $FM_l.TAG_{n-1} = minNET_{n-1}$, thus FM_l needs to advance to its current logic time again. Since $Lfed.lookahead = 0$, and $Lfed$ advances its time using $TARA$, the HLA rules allow $FM_l.TARA(current\ time + zero\ lookahead).minNET_n > minNET_{n-1}$: FM_l needs to advance time according to the minimal of $Ufeds'$ TAG times. From (5.2), $FM_l.TAG_{n-1} < min(FM_u[i].TAG_n, \forall FM_u[i] \in FMSet_u) = minNET_n$.Thus $FM_l.TARA(minNET_n)$ will request time with a greater value than the current time of FM_l, which absolutely obeys the HLA rules.

From the above reasoning, we assure that gateway federates keep strict compliance with the HLA rules in advancing time and generating TSO events.

5.4.2.2 Deadlock Free

Next, we show that the synchronization algorithm will not cause deadlock, by proving that all federates, that is, FM_u, FM_l and simulation federates, are able to obtain grants to time advance requests within a simulation's end time. Let T_n denote the time to which a federate module advances at the nth iteration of the algorithm.

$\forall FM_u[i] \in FMSet_u, FM_u[i].TARA(T_n)$. Let $NeighborFMSet_i$ be the set of federates joining the same gateway federation with $FM_u[i]$. Obviously, in a set-based federation community (see Figure 5.3(B)), each of these federates is an upper federate module hosted by an individual gateway servicing a unique user federation.

$\forall RemoteFM[k] \in NeighborFMSet_i, RemoteFM[k] \neq FM_u[i]$, let $UTimeAdv[k]$ be the time that $RemoteFM[k]$ have recently requested on the initiation of $FM_u[i].TARA(T_n)$. Here we use $min(UTimeAdv)$ to denote the minimal value of all the requested time by the federates in $NeighborFMSet_i$. There are two possibilities, $T_n = min(UTimeAdv)$: According to the HLA rules, $FM_u[i]$ will be granted to T_n. Referring to the previous proof on the correctness of upper federate modules, we have $T_n > FM_u[i].TAG_{n-1}$. This means $FM_u[i]$ can advance its time without being blocked.

$T_n > Min(UTimeAdv)$: $\exists RemoteFM[s] \in NeighborFMSet_i$, and $UTimeAdv[s] = min(UTimeAdv)$, then $RemoteFM[s]$ is able to be granted to $UTimeAdv[s]$ immediately. When other FMs in the same remote gateway federate in which $RemoteFM[s]$ resides get $TAG = UTimeAdv[s]$, the min-$NextEventTime$ in the corresponding remote user federation will increase due to the continuous attempts of time advance made by the simulation federates. Thus the $min(UTimeAdv)$ of $NeighborFMSet_i$ will increase too. With this procedure repeating, eventually we will have $T_n = min(UTimeAdv)$. Therefore, the condition for the RTI to grant $FM_u[i]$ to T_n is satisfied at this stage.

From the above, we can see that $FM_u[i]$ is able to advance time continuously within the simulation termination time under any condition.

When $FM_l.TARA(minNET_n)$ is initiated, according to the definition, we have

$$minNET_n \leq \text{the current } LBTS \text{ of } FM_l \qquad (5.7)$$

(5.7) implies that FM_l will be granted $minNET_n$ immediately. We consider the following two cases. $minNET_n = minNET_{n-1} = T$: With FM_l issuing $TARA(T)$ repeatedly, all the TSO events with timestamp T will sooner or later be delivered to FM_l at some point. On that point, FM_l will detect $minNET_n > T$.

$minNET_n > minNET_{n-1}$: This means that FM_l progresses to a greater time. With the advancing of FM_l's logic time, the user federates will also definitely get time granted within the simulation termination time.

In the case that the federation community is not set based, $NeighborFMSet_i$ may contain only one remote lower federate module ($ReFM_l$) (belong to remote gateway X). With the remote federations linked through gateway X keeping advancing time, the upper federate modules will eventually be granted to the time $ReFM_l$ passed to them, $min(UTimeAdv)$. $ReFM_l$ will be granted in the gateway federation formed by $NeighborFMSet_i$; gradually other $RemoteFM$s will be granted and the $min(UTimeAdv)$ increased. From the above we can conclude that all types of federates involved in the federation community are able to progress time under the management of the algorithm.

5.4.2.3   Correct TSO Event Transmissions

This issue concerns the bi-directional propagation of events, that is, from the local federation to the remote federations and the other way around. From the local federation to the remote federations, FM_l collects TSO events from local user federation. The algorithm keeps FM_l issuing $TARA$ to the current *minNextEventTime* unless an increase in the value has been detected. This guarantees that FM_l has retrieved all local events with timestamp equal to *minNextEventTime* prior to the next $TARA$ invocations by FM_us.

From the remote federations to the local federation, let us consider a simulation federate's (SF) using method TAR to advance time. Given $SF.TAR(T_n)$ in the nth time advance request, and SF is granted to T_n, $SF.TAG_n = T_n$.

According to the condition to satisfy this time advance request defined in the HLA rules, we have

$$SF.LBTS_n > T_n \qquad\qquad (5.8)$$

As FM_l is time regulating, if $FM_l.TARA(T_x)$ is invoked at this point, and T_xT_n: Since $SF.LBTS_n$ should not be greater than the time requested by FM_l, we have $SF.LBTS_nT_xT_n$, this conflicts with in equation (5.8)

We can conclude that FM_l must have initiated $TARA(\tau)$, with $\tau > T_n$ (5.9) From (5.9) and the algorithm, we have $\forall FM_u[i] \in FMSet_u$: $FM_u[i].TAG_n = \tau$, which means all FM_us have already issued $TARA(\tau)$ and been granted. This also means that FM_us have generated all events with *timestamp* $< \tau$ in the gateway federation. Let's refer to the subsection on "deadlock free", $\forall RemoteFM[k] \in NeighborFMSet_i$, from (5.8) and (5.9), we have $RemoteFM[k]$ must have performed $TARA(\tau + \delta), \delta \geqslant 0$ (5.10) Let $RemoteMinNET[k]$ denote be the current *minNextEventTime* of the remote simulaiton federation for which $RemoteFM[k]$ serves. From (5.10), we have

$$RemoteMinNET[k] = \tau + \delta \qquad \geq \tau > T_n \qquad (5.11)$$

We can conclude from (5.11) and the above reasoning that $RemoteFM[k]$ had processed all events with *timestamp* $\leqslant \tau + \delta$ from the remote simulation.

$\forall FM_u[i].TAG_n = \tau$ also means: (1) $\forall FM_u[i] \in FMSet_u$, $FM_u[i]$ had put all remote TSO events with *timestamp* $< \tau$ into buffer and (2) FM_l had simulated all the remote TSO events with *timestamp* $< \tau$ in the local simulation federation prior to its time advance request. The same facts apply to any other related remote federation of the federate gateway. Thus, $SF.TAG_n = T_n$ means that SF had received all remote events with timestamp less than or equal to Tn from all simulation federations linked to the gateway federate.

Although the above proof is given from the perspective of a gateway federate associated with a user federation, the proof can be generalized to the gateway federates servicing only gateway federations. The correctness of the second algorithm can be also proved easily in a similar manner.

5.5 Experiments and Results

In order to verify the correctness and investigate the overhead incurred in the proposed synchronization mechanism and its scalability, we carried out a number of experiments to benchmark the efficiency of synchronization in federation communities (multiple-layer federation community, peer-to-peer federation community, and Grid-enabled federation community) in both LAN environments and a WAN environment between the United Kindom (Birmingham) and Singapore.

The DMSO RTI NG 1.3 with Java bindings and its associated time advance benchmark application are used in the experiments. The identical participating federates are both time constrained and time regulating, and each attempts to advance its time from 0 to 10,000 with both time-step and lookahead set to 1.0. A time synchronization benchmark records the number of RTI time-step cycles that can be processed by the underlying networked RTIs per second. A time-step cycle starts at the point where a benchmark federate issues a time advance request and ends at the point where the request is being granted. The results are reported as the number of time advances granted per second (TAGs/second). The experimental results presented in this section are averaged from the outputs of a number of runs, and the deviations are very insignificant.

In these experiments, a federation community has been treated from the simulation federates' point of view rather than the underlying RTIs, with the underlying RTIs in a federation community operating as a whole "composite" infrastructure. No matter how the participating federations (or RTIs) are interconnected, those RTIs remain transparent to the simulation federates (this is exactly the case for a common federation). The TAG rate is measured from the federates' point of view.

5.5.1 Experiments on Multiple-Layer Federation Community Networks

In [43], we investigated the synchronization efficiency of two types of multiple-layer federation community networks (see Figure 5.3), that is, the hierarchical federation community network (HFC) and the set-based federation community network (SFC). The preliminary investigation focused on the two alternative architectures' scalability. In the context of multiple-layer federation communities, a federation community may scale either vertically, such as by increasing the number of hierarchical layers of the structure in an HFC, or horizontally, such as by increasing the number of gateway federations that a gateway federate joins in an SFC. Figure 5.10 and Figure 5.11 present the scenarios of the benchmarks for SFC and HFC, respectively. The benchmark federates involved in each scenario are enclosed in the corresponding indexed circle.

The experiments were conducted in a LAN environment consisting of a number of PCs. Each PC has 256M bytes of memory and one single Intel CPU (at either 1.4GHz or 1.7GHz) while connectivity is provided by a 100Mbit Ethernet switch. We first performed a synchronization benchmark using a flat federation, and the set of results (Figure 5.12) are used as a standard reference for the remaining benchmarks on multiple-layer federation communities. The benchmark results for SFC and HFC are presented in Figure 5.13.

In the flat federation, the TAG rate remains about 20 times per second with the number of federates varied from two to eight. With two federates, the TAG rate measured on federation community is 7.8 TAGs/second, about 40% of the corresponding flat federation result. The TAG rates of both SFC and HFC decrease with the number of federates. All simulation federates report almost the same TAG rates, as they are designed to be regulated by each other and our synchronization mechanism has properly enabled this in the final federation communities regardless of their architectures (the same applies to all other experiments introduced in Sections 5.2 and 5.3).

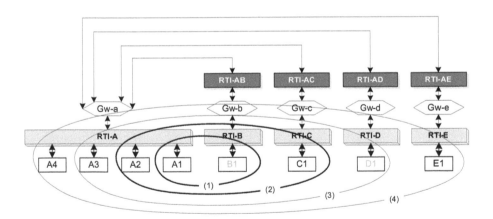

FIGURE 5.10
Synchronization benchmark scenarios for set-based federation community networks.

Obviously, the SFC's synchronization efficiency is superior to that of the HFC. With more levels in HFC, it takes more intermediate stages for the time synchronization to propagate through the gateway federates to the remote benchmark federates as compared to SFC. From Figure 5.13 it can be

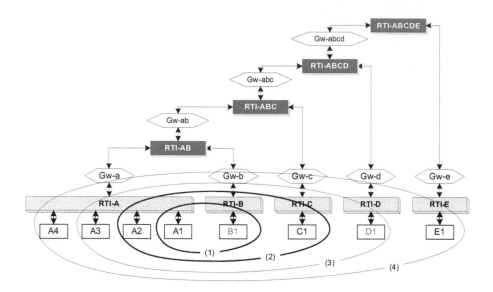

FIGURE 5.11
Synchronization benchmark scenarios for hierarchical federation community networks.

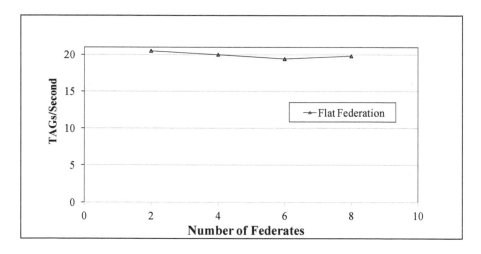

FIGURE 5.12
Synchronization benchmark results for a flat federation over a low-performance platform.

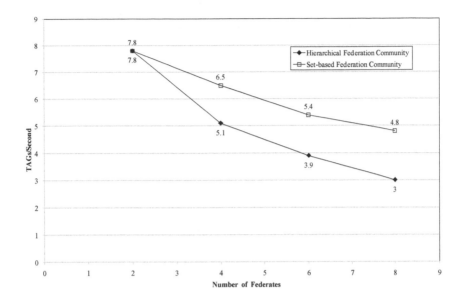

FIGURE 5.13
Synchronization benchmark results for hierarchical federation community and
set-based federation community.

observed that the lowering of the TAG rate (about 20% of the corresponding flat federation result) starts to bottom out for SFC, and this indicates
that SFC has potentially good scalability. As far as synchronization efficiency
is concerned, the results suggest that scaling horizontally tends to result in
better performance than scaling vertically when constructing a multiple-layer
federation community.

5.5.2 Experiments on Peer-to-Peer Federation Community Networks

This series of experiments focuses on the scalability study of peer-to-peer federation community (See Figure 5.3(A)). These experiments aim to (1) compare
the performance of peer-to-peer federation community and flat federation, and
(2) investigate the scalability with respect of the number of federations in a
peer-to-peer federation community. The experiments were based on a high
performance PC cluster at Birmingham/UK consisting of one master node
and 54 worker nodes. Each node has 2G bytes of memory and two Intel Xeon
3GHz processors, while connectivity is provided by a Gigabit Ethernet Switch.
Each benchmark federate occupies an individual work node separately.

Figure 5.14 presents the synchronization benchmark results using a flat federation. In the time advancement benchmark, the TAG rate decreases with the number of federates. The rate decreases less rapidly when the number of federates is greater than 6. The TAG rate is 2279 times per second for two federates down to about 373 times per second (about 16% of the case of two federates) for twenty four federates.

The benchmark federates were then partitioned into two separate federations evenly (one federation will contain one more federate than the other when the total number of federates is odd) and formed a simple peer-to-peer federation community using a gateway federate on an additional worker node. The experimental results are also presented in Figure 5.14. The rate decreases slightly with the number of federates in contrast to the flat federation case. The TAG rate is 525 times per second (about 20% of the corresponding flat federation result) for two federates down to about 287 times per second (about 55% of the case of two federates and about 77% of the corresponding flat federation result) for twenty four federates. The synchronization efficiency of the peer-to-peer federation community tends towards that of the flat federation and the overhead introduced by synchronization becomes more and more ignorable with the increasing number of federates. The encouraging results indicate that the peer-to-peer community scales very well when the number for federations is limited.

In another set of experiments, we fix the total number (24) of federates and make them evenly join a number (2, 3, 4, 6, and 8) of federations. The TAG rates in these scenarios are reported in Figure 5.15. The TAG rate is 287 times per second for two federations down to about 96 times per second (about 33% of the case of two federations) for eight federations.

This suggests that the number of federations in a community should be limited in the case of requiring extremely high synchronization efficiency. Another observation is that the TAG rate decreases very slightly when the number of federations exceeds four. The results also indicate that the scalability of the synchronization mechanism is good in terms of the number of federations.

It is pertinent to note that the overhead incurred by synchronization is only part of the load of a federated simulation system, which also includes the models' computational load and communication load for message deliveries. When the models are highly complex (this is often the case for federation communities), the relative contribution of the synchronization overhead to the overall system load is decreased. The same is true as the number of federates increases.

5.5.3   Experiments on Grid-Enabled Federation Community Networks

The last set of experiments aims to investigate the correctness and performance of the synchronization mechanism in Grid-enabled federation community. A high-performance cluster at Singapore has been used in addition to

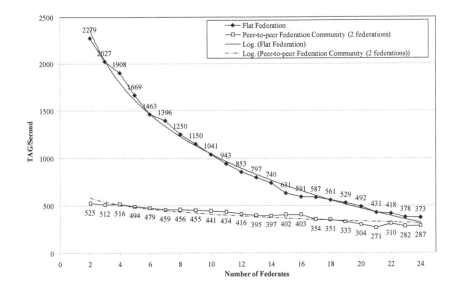

FIGURE 5.14
Synchronization benchmark results for a flat federation versus a simple peer-to-peer federation.

the cluster at Birmingham. The cluster at Singapore consists of one master node and fifteen worker nodes, and each node has 4G bytes of memory and two Intel Dual Core 3GHz processors while connectivity is provided by a Gigabit Ethernet Switch. The firewall rule on each cluster only allows external authorized users to access the master node via a designated port.

The configuration of the testbed is illustrated in Figure 5.16. Grid services provided by GT4.1.2 have been enabled to the master nodes of both clusters. In order to initiate the Grid-enabled gateway, a *ClientFederateAmbassador* service should be started to host the gateway thread on the master node at Birmingham, namely "Gw-b," and a *ProxyRTIAmbassador* service should also be started to host the corresponding proxy, namely "Proxy b."

Each benchmark federate occupies an individual work node separately. The benchmark starts with one federate at Singapore cluster and another at the Birmingham cluster. We then increased the number of federates equally until twelve federates were involved at each side (denoted as A1 to A12 at Singapore and B1 to B12 at Birmingham). After that, we only add federates (denoted as B13 to B24) at the Birmingham cluster to a total number of thirty-six federates. The setup of the Grid-enabled federation community follows the architecture as shown in Figure 5.5(A). Proxy b directly interacts with federates B1 to B24 through an RTI session at Birmingham while Gw-b

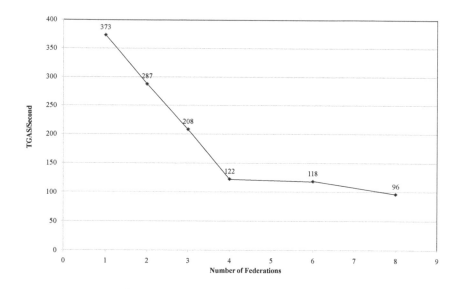

FIGURE 5.15
Synchronization benchmark results for peer-to-peer federation communities over a high-performance platform.

interacts with federates A1 to A12 through another RTI session at Singapore. A Grid-enabled gateway between the two RTI sessions were formed by gluing Gw-b together with Proxy b through Grid invocations over the Grid. Figure 5.17 presents the benchmark results.

The TAG rate is 1.61 times per second for two federates down to about 1.48 times per second for four federates. The TAG rate remains almost constant when the number of federates is less than twenty-six. With more federates added at Birmingham side, the TAG rate decreases slightly down to 1.15 times per second with thirty-six federates in total. The results exhibit an encouraging scalability of the synchronization mechanism upon the Grid-enabled federation community.

The experimental results are mainly determined by the communication overhead between the clusters. We have measured the times required for RTI service invocations (*queryMinNextEventTime* and *timeAdvanceRequestAvailable*) across the Grid, and each of them takes about 300 milliseconds to complete. The two calls will be invoked at least once in order for the gateway to advance the time at both sides (see Figures 5.8 and 5.9). Even if the overhead of other operations can be ignored, the interval for the gateway to successfully advance its simulation time once is at least 600 milliseconds. Because only when the gateway advances its time successfully can other federates pro-

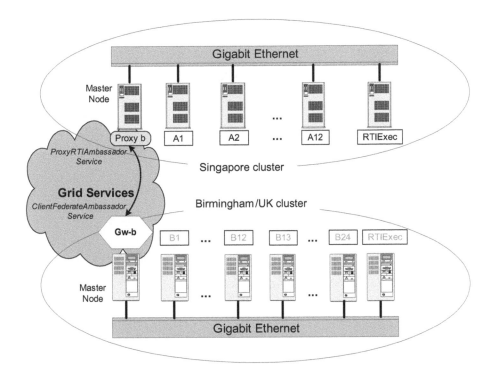

FIGURE 5.16
Testbed for benchmark of Grid-enabled federation community over WAN.

ceed with a new time advance request, the TAG rate of a federate should not exceed about $1/(0.6) = 1.66$ times per second. This figure is in accordance with the experimental results. Furthermore, due to the delay caused by Grid invocations at the gateway, the overhead involved in the intra-federation synchronization is well hidden.

In addition to the outputs described above, the benchmark federates in all experiments presented in this section can advance their times and get grants properly in exactly the same manner as in a flat federation. This observation further indicates the correctness and user transparency of the proposed synchronization mechanism.

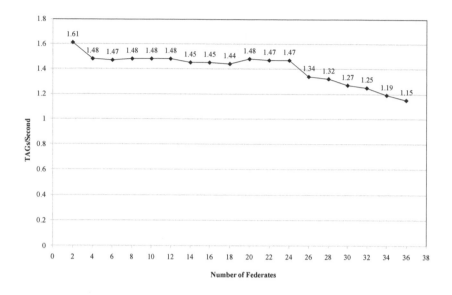

FIGURE 5.17
Synchronization benchmark results for Grid-enabled federation community.

5.6 Summary

This chapter is concerned with issues related to synchronization in federation community networks with alternative approaches for constructing federation communities and various architectures of federation community networks introduced. The chapter presented a mechanism for supporting synchronization crossing the boundaries of interlinked federations. We have investigated the issues and design of two generic synchronization algorithms. Based upon the gateway federate approach, the synchronization mechanism was developed to properly propagate time constraints of each individual federation all over the network, thus to coordinate the time advances of all simulation federates and TSO event deliveries among them. This has been achieved without the risk of deadlock or violation to the HLA rules.

All the objectives set at the beginning of this chapter have been fulfilled. A mathematical proof as to the correctness of the synchronization algorithms has been provided. Our design supports the reuse of legacy federates at both the executive level and code level while enabling synchronization among federations. It avoids developers' effort for reengineering simulation federates in order to make use of the proposed synchronization mechanism. The synchronization mechanism is platform neutral and model independent, and it has been successfully applied in federation communities of a variety of architectures.

The Grid-enabled federation communities have been successfully constructed with time management fostered by our synchronization algorithms. An individual federation operating inside an administrative domain has been presented as Grid services using GT4.1.2 and can be by easily accessed for use by external users.

A number of experiments have been performed to investigate the correctness and benchmark the performance of the synchronization mechanism. The standard benchmark applications available in the DMSO RTI software suite (1.3 NG Version 6) have been adopted to benchmark the synchronization mechanism over LAN and WAN environments in terms of efficiency of time advance and scalability. The experimental results on multiple-level federation community networks over LAN indicate the mechanism scales better in a set-based federation community than in a hierarchical federation community, and it suggests that scaling horizontally is preferred when constructing a multiple-layer federation community. The experimental results on peer-to-peer federation community networks show that the synchronization mechanism provides encouraging synchronization performance, especially in the case of large-scale distributed simulation. It can be observed that with the number of federates increasing the federation community powered by the proposed algorithm can reach about 77% performance of the corresponding flat federation in terms of synchronization. The synchronization performance on Grid-enabled federation communities is limited by the communication overhead over the Grid. Nevertheless, the synchronization mechanism scales quite well. The experimental results imply that the mechanism suits compute-intensive and large-scale simulation when synchronization among federates is not frequent.

Part III

Evaluation of Alternative Scenarios

6

Theory and Issues in Distributed Simulation Cloning

CONTENTS

This chapter presents the theory of distributed simulation cloning and the critical research issues involved. It defines basic concepts and notations that are used in the technology. Different types of federate cloning and scenario cloning in distributed simulations are identified and classified.

6.1   Decision Points

As discussed in Section 3.2, during the execution of a simulation, a federate may face different choices to perform alternative actions. The federate is cloned at such **Decision points** according to some predefined triggering conditions or rules. A decision point represents the location in the execution path where the states of the system start to diverge in a cloning-enabled simulation. "Cloning" differs from simple replication in the sense that Clones of the original federate execute in different paths rather than simply repeat the same executions, even though the computation of clones is identical at the decision point. From the decision point onward, a simulation spawns multiple execution paths to exploit alternative scenarios concurrently.

From the user's point of view, Decision points can be specified either at the modeling stage or at runtime. A user may predefine the Cloning triggers for a decision point, which comprise the cloning conditions, alternative actions, and properties of the trigger. The properties of a trigger describe how it is defined: This may be based on (1) the values of an object instance, (2) the status of the model's internal states, (3) a certain simulation time, or (4) dependency on other triggers. At the modeling stage, the user can specify the conditions of the Decision points and the alternative actions for the Decision points. During runtime, the user can insert Decision points dynamically and can even specify alternative execution paths at a decision point interactively. The cloning trigger will initiate cloning when necessary.

Any occurrence of cloning has an impact on the existing scenarios and other federates in a distributed simulation. Section 6.2 classifies the simulation cloning from an individual federate's point of view, whereas Section 6.3 defines different cloning mechanisms from a scenario's perspective.

6.2   Active and Passive Cloning of Federates

When a federate reaches a decision point, it makes clones on its own initiative. This federate is said to perform active cloning. Each clone of the federate executes a separate scenario. An **active cloning** results in the creation of new scenarios.

In distributed simulations, there are multiple federates interoperating with each other. When one federate splits into different executions, the partners who interact with this federate may have to spawn clones to perform proper interaction even though the partners have not yet met a decision point. Those partners are said to perform passive cloning. Generally speaking, the clones generated in an active cloning have separate initial states while those created in a **passive cloning** have identical initial states. Only the active cloning leads to the creation of new scenarios, and this induces passive cloning in some other federates.

6.3   Entire versus Incremental Cloning

A simple approach to keep the correctness of the simulation is to clone the whole simulation whenever any federate reaches a decision point, that is. **entire cloning**. Each cloned simulation consists of a separate set of federates that reports results independently. One can see that the scalability of distributed simulation is a challenge in this case. Another approach replicates the simulation incrementally; that is, only those federates whose states will alter at a decision point need to be cloned, other federates will remain intact. This alternative approach is referred to as **incremental cloning**. Such an incremental cloning approach shares computation between federates in alternative scenarios, and provides a more efficient and scalable method to clone the distributed simulation.

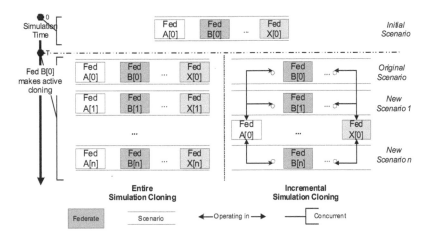

FIGURE 6.1
Entire cloning versus incremental cloning.

6.3.1 Shared Clones

Figure 6.1 illustrates the different effects of an active cloning using entire cloning versus incremental cloning. For those federates whose states are not affected, the incremental cloning mechanism allows them to operate in the new scenarios in addition to the original one as **shared federates** (clones), such as *Fed A[0]* and *Fed X[0]* in the right half of Figure 6.1. A shared clone may subsequently perform cloning passively during the execution of the simulation on demand of its partners.

Normal clones are those clones that operate in a single scenario (e.g. *Fed B[i]* in Figure 6.1); this term is used in this book to distinguish them from the shared clones. The clones created from the same root federate are referred to as **sibling clones**. The federate being cloned is referred to as the **parent federate** (clone) of the clones directly replicated from it. Those federates that interact within the same scenario are known as **partner federates**[1].

A clone needs to inherit the same RTI objects from the parent federate. We name the object instances registered by the original federates prior to cloning as **original object instances** whereas we use **image object instances** to denote those object instances re-registered by the clones in the state recovery procedure.

---

[1] In our design, each federate (clone) is associated with a scenario ID, from which partnership can be distinguished among different federates (clones).

6.3.2　Theory and Issues in Incremental Cloning

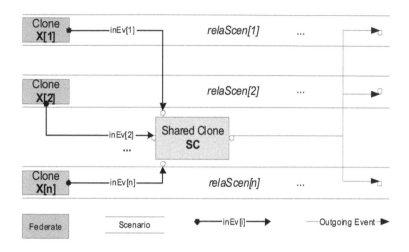

FIGURE 6.2
A typical shared clone.

The incremental cloning mechanism enables a shared clone to execute in multiple scenarios as long as it keeps receiving identical events from corresponding federates in all scenarios in which it participates. This design aims to avoid repeating identical computation among scenarios as much as possible. The shared clone persists in this mode until the condition for triggering passive cloning is met.

A typical shared clone is shown in Figure 6.2. The shared clone (SC) executes in n concurrent scenarios, and those scenarios are said to be SC's **related scenarios** (written as $RELASCEN = \{relaScen[i]|i = 1, 2,...,n\}$). Let $X = \{x[i]|i = 1, 2,...,n\}$ denote the set of sibling clones that are created from the same simulation federate x, with $x[i]$ operating in $relaScen[i]$. SC may receive events from $x[i]$ and generate events for each related scenario. It is unnecessary to perform extra checking on the events generated by SC, as those events must be identical in any scenario. However, the events received by SC must be checked.

Definition 1 (Sensitive Update) If an object instance $ObjX$ registered by federate has been discovered by SC, then SC treats $ObjX$ and its image objects (see Section 6.3.1) as a set of sensitive object instances. Obviously, the object class to which $ObjX$ belongs must be published by x and subscribed by SC. Let $inEv[i]$ represent an update of $ObjX$ (or its image objects) issued by any $x[i] \in X$; then $inEv[i]$ is defined as a **sensitive update** for the shared clone SC.

Definition 2 (Sensitive Interaction) Any interaction class published by x and subscribed by the shared clone SC is regarded as a sensitive interaction class. Let $inEv[i]$ represent an interaction of any sensitive interaction class sent by any $x[i] \in X$; then $inEv[i]$ is defined as a **sensitive interaction** for the shared clone SC.

A **sensitive event** is defined as a sensitive update or interaction. A shared clone may present nonsensitive events straightforwardly to its simulation model without extra checking, whereas it has to check each sensitive event before conveying it to the simulation model. A non-sensitive event can be an event sent by another shared clone executing in all related scenarios of the receiver. A sensitive event must be compared with corresponding counterpart events. In each round of event comparison, the first received sensitive event is referred to as the **target event** by subsequent counterpart events.

Definition 3 (Comparable Updates) Any two sensitive updates for a shared clone are **comparable** to each other only when the following conditions are satisfied:

- *They carry equivalent timestamps.*

- *They are updates of two individual image objects (or an original object and one of its image objects) representing the same original object.*

.

Definition 4 (Comparable Interactions) Any two sensitive interactions are comparable only when the following conditions are satisfied:

- *They carry equivalent timestamps.*

- *They belong to the same sensitive interaction class.*

- *They originate from two individual sibling clones.*

A shared clone should not compare received interactions that are not sent by sibling clones even if they belong to the same interaction class. According to Definition 3 and the definition of original and image object instance, it is obvious that comparable events must originate from sibling clones.

Definition 5 (Identical Events) Comparable events are called **identical** if they have the same associated attributes/parameters and the values of all attributes/parameters are identical.

Comparable events must be checked to verify whether or not they are identical. If a shared clone detects that any two comparable events are not identical, the shared clone has to perform passive cloning to handle this situation. On the other hand, the shared clone may remain intact if

- *All received comparable events are identical.*

- *The shared clone receives comparable events from all the sibling clones in the related scenarios before it is granted a simulation time greater than (or equal to) the target event's timestamp.*

If either condition is not met, it means that the shared clone has obtained different behaviors from related scenarios and requires passive cloning. As a consequence, the federate previously shared and its clones created in this passive cloning each operate as a normal clone in only one individual scenario (at least until the next decision point). Figure 6.3 depicts the relationship among the terms defined in this section.

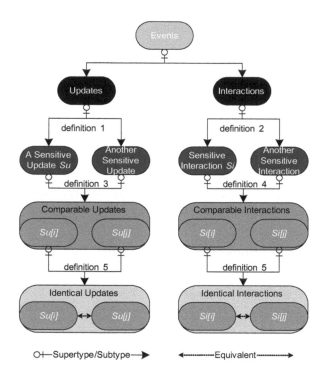

FIGURE 6.3
Relationship between terms related to shared clones.

6.4   Scenario Tree

In the context of distributed simulation cloning, each clone (federate) is an individual entity, whereas each scenario is a dynamic group, which involves a changing combination of member clones. Each scenario reports simulation results independently; it is the basic unit in our consideration and discussion. Only active cloning can drive the creation of new scenarios. We utilize a tree data structure to represent the relationship and development of the scenarios.

Figure 6.4 gives an example of a cloning-enabled simulation in which *Fed A[0]* operates as an event publisher, and *Fed B[0]* and *Fed C[0]* exchange events. Part (A) illustrates the details of the overall cloning procedure. Part (B) gives an abstraction of the scenario tree, in which each **parent** node (full circles) represents an occurrence of active cloning. The **leaf** nodes (empty circle) stand for active scenarios at the current simulation time. An active cloning results in spawning new scenarios; this is reflected in the figure as parent nodes that have multiple children. Each scenario is marked as *S[i]* (*i* = 0, 1, 2, ...). The scenario tree grows along the simulation time axis as follows:

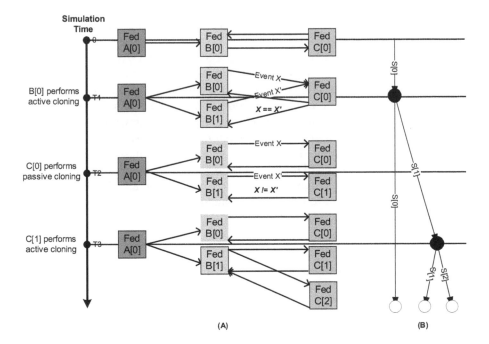

FIGURE 6.4
Example of incremental cloning and scenario tree.

- At simulation time *0*, there exists a single scenario *S[0]*. When simulation time is advanced to T_1, *Fed B* meets a decision point and performs active cloning, splitting into clones *B[0]* and *B[1]*. A new branch *S[1]* is created in the scenario tree at this point. An event generated by *B[0]* is named as event X while an event from *B[1]* is called event *X'*. *Fed C[0]* keeps intact and this will continue as long as events X and *X'* remain the same. *Fed C[0]* operates as a shared clone for the duration between time T_1 and time T_2.

- At simulation time T_2, an event X deviates from event X'; this incurs a passive cloning of *Fed C[0]* and results in the birth of clone *C[1]*. This passive cloning does not trigger any change in the scenario tree.

- At simulation time T_3, *Fed C[1]* performs an active cloning, spawning off clone *C[2]*. A new scenario is created and marked as *S[2]* in the scenario tree.

A Combinatorial explosion of scenarios in distributed simulation cloning may occur in some situations. The number of possible scenarios is determined by (1) the number of active cloning federates, (2) the times those federates perform active cloning, and (3) the candidate choices that each decision point represents. Human intervention may reduce the Combinatorial explosion, but it is difficult to reach a general solution [5].

In practice, it is unlikely that such a Combinatorial explosion will occur. For example, in a supply chain simulation, one company may wish to examine its own decision strategies concurrently. It is unlikely that one company could manipulate its partners' internal decision policies. So in this case, there will be only one or a few federates that perform active cloning while the remaining federates perform passive cloning at the request of the active cloning federates.

From the above discussion we observe that different scenarios may have common member clones. The relationship among scenarios and Clones can be complex and highly dynamic. There is a need for an identification and partitioning mechanism to manage the concurrent scenarios in the distributed simulation cloning procedure. This requires us to

- Represent the relationship among scenarios.

- Identify each scenario and clone to enable control.

- Partition the event messages belonging to different scenarios.

- Support the sharing of Clones between scenarios.

- Provide reusability to user federates while enabling the former functionalities.

In Chapter 8 we will address the above problems by presenting a recursive region division solution and a point region solution to manage the scenarios using Data Distribution Management (DDM).

6.5   Summary

This chapter gave the basic theory and identifies issues involved in distributed simulation cloning. Different types of cloning were introduced, such as active and passive cloning, and total and incremental cloning. Key terms were defined, which include decision point, normal clone and shared clone, and partner federate and sibling Clones. Although some concepts, for example decision point, have been identified in [60] for parallel simulation cloning, our book is the first to provide a systematic definition of simulation cloning at both the federate and distributed simulation level.

This chapter also described the theory of incremental cloning. Sensitive events and related concepts were explained. Similar ideas for incremental cloning have subsequently been adopted in the parallel simulation domain [59].

A scenario tree was designed to represent the evolvement of concurrent scenarios resulting from cloning and the dynamic relationship among them. This chapter clearly described the interdependencies between scenarios, Clones, RTI objects and events which is missing in previous approaches to cloning of HLA-based simulations.

7

Alternative Solutions for Cloning in HLA-Based Distributed Simulation

CONTENTS

Simulation cloning is designed to satisfy the requirement of examining alternative scenarios concurrently. In this chapter, alternative solutions are proposed and compared from both the qualitative and quantitative points of view. In terms of federation organization, candidate solutions can be classified as either single federation or multiple federation. In order to guarantee the correctness and optimize the performance of the whole cloning-enabled distributed simulation, the single federation solution requires an additional mechanism to isolate the interactions among alternative executions. Data Distribution Management (DDM) is one of the candidate approaches. To measure the trade-off between Complexity and efficiency, we also introduce a series of experiments to benchmark various solutions at the RTI level.

7.1 Single-Federation Solution versus Multiple-Federations Solution

When a federate is cloned, we can create multiple federations to meet the demand of executing alternative scenarios or generate new federates to operate in the original federation without intervening in the execution of any other scenario. This book uses Multiple-Federation (MF)Solution to denote the former design, and Single-Federation (SF) Solution to denote the latter one.

Figure 7.1 depicts the cloning of a simulation using both solutions. Federates inside the dashed rectangle represent the clones originating from a common ancestor. Following the cloning action that is triggered at the decision point, both original federates (A and B) duplicate themselves to form two different scenarios. By applying the SF solution, the new federates *Fed A[1]* and *B[1]*participate in the original federation *RTI[0]* and there is no need for an additional federation to support the new scenario (labeled). By applying the MF solution, *Fed A[1]* and *B[1]*form another federation *RTI[1]*to facilitate another scenario (labeled).

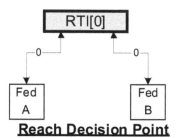

FIGURE 7.1
Example of single-federation solution and multiple-federation solution.

The two solutions mentioned above involve different research issues and problems, especially at the RTI level. Table 7.1 gives a comparison showing

TABLE 7.1

Comparison between Single-Federation and Multiple-Federations Solutions

Issues	Single-Federation	Multiple-Federation
Interaction	Additional mechanism is needed to deal with unnecessary event-crossing among concurrent scenarios	Interaction between clones in various scenarios isolated by default
Synchronization	Unnecessary synchronization among clones in different scenarios is inevitable	Synchronization among clones in different scenarios is eliminated
Complexity of Sharing	Clone sharing is available in a single federation	Clone sharing is difficult among federations
Robustness	If an RTI instance crashes, the simulation will fail	Multiple RTI instances for one user federation are maintained; thus one RTI instance crash will not result in failure of the whole simulation
Management of Clones	Management of clones is easier inside the same federation	Management of clones crossing multiple federations is indirect

the advantages and disadvantages of both solutions from different viewpoints. To make a trade-off among these issues is difficult. As RTI does not provide destination-specific delivery in its Object Management services, it is mandatory that the Single-Federation solution requires an additional mechanism to isolate interactions among clones in different scenarios.

7.2   DDM versus Non-DDM in Single-Federation Solution

To isolate separate scenarios, a straightforward approach is to filter events at the receiver side. Each event is attached with the exclusive identity of the scenario of the sender. The receivers discard those events from other scenarios and merely reflect those belonging to the same scenario. Minimal effort is required to enable this filtering in addition to the standard RTI services.

It is also possible to use Data Distribution Management (DDM) services to partition scenarios in the overall cloning-enabled distributed simulation. In general, the purpose of DDM services is to reduce the transmission and receipt of irrelevant data by the federates. DDM services are employed by the federates, which can be interpreted as data producers and consumers, to assert properties of their data or to specify their data requirements respectively based on specified regions. The RTI then distributes the data from the producers to the consumers based on the match between the properties and the requirements. DDM controls the efficient routing of class attributes and interactions via the RTI.

Routing spaces are a collection of dimensions that represents coordinate axes of the federation problem space with a bounded range [86]. A region defines a multidimensional sub-space in the routing space by defining the lower bound and upper bound on each dimension of the routing space. Figure 7.2 illustrates an example of routing space and regions, in which two dimensions "X" and "Y" define a routing space. Three regions R_1, R_2, R_3 are indicated as the large rectangles in the figure. R_1 overlaps with R_2 in area O_1 and R_2 overlaps with R_3 in area O_2. Obviously, there is no overlap between R1 and R_3.

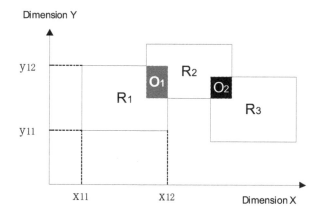

FIGURE 7.2
Example of routing space and regions.

In DDM, data producers and consumers specify their data properties and data requirements by providing update regions and subscription regions. Data connection will be established between a pair of federates only when an update region and a subscription region overlap. With this property, DDM seems to be a natural candidate for developing the mechanism to restrict the interaction among clones to within the same scenario.

The RTI offers a region modification method to enable the dynamic change of sub-space without creating a different region. The change takes effect immediately after notifying the RTI. This feature enables the adjusting of a clone's "characteristic" region at runtime; thus dynamic routing and filtering of events from one clone to different scenario combinations can be realized.

The RTI specification leaves the DDM implementation details to the RTI implementers. The current DMSO RTI-NG supports a number of DDM data filtering strategies, one of which is the **StaticGridPartitioned** strategy. This strategy partitions individual spaces into a grid in which each grid cell is assigned a separate reliable and best-effort channel [63]. Wise use of the Static-GridPartitioned strategy can increase performance through sender-side filtering.

By assigning a scenario-specific region to one set of clones, the interactions will automatically be confined to this scenario. However, this incurs some extra overhead for managing the regions and increases the Complexity of implementation. It is not necessary to introduce the DDM mechanism to the multiple-federation solution, as the communication traffic has already been confined within each federation. Thus, there are three candidate solutions for cloning HLA-based distributed simulations: Multiple-Federation solution (**MF**), DDM Single-Federation solution (**DSF**), and Non-DDM Single-Federation solution (**NDSF**). In order to evaluate these solutions, another important criterion is their efficiency. The trade-off between efficiency and Complexity is a main concern of the distributed system designer. Section 7.4 will introduce the benchmark experiments to measure the overall performance of the three solutions in terms of execution time.

7.3   Middleware Approach

No matter what kind of solution is chosen to perform simulation cloning, a critical principle is the reusability of the user's program. Our solutions should minimize the modification to the user's existing code even though it is difficult to eliminate all extra Complexity when enabling simulation cloning. We propose a middleware approach to hide the Complexity as indicated in Figure 7.3.

We extend the standard RTI to RTI++ to encapsulate cloning operations directly related to the RTI while presenting the standard RTI interface to the user. The user code still uses the standard RTI interface while the enhanced functionality remains transparent to the user. The middleware sits between the user's code and the real RTI, and contains a library for cloning management and the RTI++.

As mentioned previously, any of the solutions discussed should work as an underlying control mechanism with transparency to federate codes. Hiding the implementation details while enabling the management of scenarios can maximize the reusability of the user's existing simulation model. Our RTI++ software is built to encapsulate the cloning-related modules while maintaining the same RTI interface to the calling federate code. Figure 7.4 gives an example of pseudo codes for implementing the above inside the middleware.

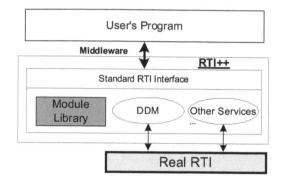

FIGURE 7.3
Example of middleware for cloning.

Get region associated with current clone federate

   CurrentRegion = *ScenManager->getCloneRegion()*;

Override Standard RTI functions with DDM enabled services

      For example:
      RTIambassadorPlus::subscribeObjectClassAttributes(theClass, attributeList, ...){
      ...

      (RTI::RTIambassador this)->subscribeObjectClassAttributesWithRegion(theClass,*
      CurrentRegion, *attributeList, ...);*
      ...
      }

   ...

FIGURE 7.4
Example of RTI++ implementation.

7.4   Benchmark Experiments and Results

The primary objective of the experiments is to provide some criteria on Complexity and efficiency to help us decide what kind of cloning method we should adopt: namely, MF, DSF, and NDSF as discussed.

   We study the Complexity and efficiency using three factors: (1) **Lower Bound Time Stamp (LBTS) computation** for time advance [1], (2) the

interaction latency between federates, and (3) **the load of the federate processes**. The LBTS computation determines the speed at which federates can advance simulation time. The interaction latency means the communication overhead for delivering events among federates. The load of the federate processes is a measure of the computational resources required for executing a simulation. The experiments explore how these crucial factors impact the overall performance. The experiments report the execution times of adopting alternative solutions in terms of the efficiency of the LBTS computation and the interaction latency. They also measure the CPU utilization of each node to study the load of federate processes under different circumstances.

7.4.1   Experiment Design

The experiments use three PCs in total (PC 1, 2, and 3 in Figures 7.5 and 7.6), in which PC 2 executes the RTIEXEC and FEDEX processes [1]. The federates that run at each independent PC are enclosed in a dashed rectangle. In our case *Fed A[i]* and *Fed B[i]* ($i \geqslant 1$) occupy PC 1 and PC 3 respectively. The PCs are interlinked via an EtherFast 100 five-port Workgroup Switch, which forms an isolated subnet to avoid fluctuation incurred by additional network traffic. The PCs' configuration is as follows:

- Intel 1700MHz Pentium IV

- 256 Mbytes of RAM

- Windows 2000 Professional

- DMSO RTI NG 1.3 V4

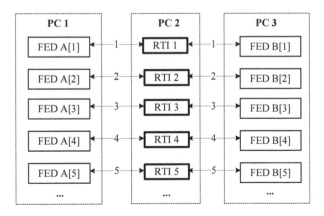

FIGURE 7.5
Test bed for MF solution.

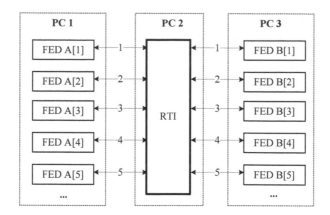

FIGURE 7.6
Test bed for DSF and NDSF solution.

The experiments emulate the simulation cloning process by increasing the number of identical federates. In Figures 7.5 and 7.6, *Fed A[1]* and *B[1]* form a pair of initial federate partners, which represent the federates to be cloned. *Fed A[i]* and *B[i](i>1)* denote the *i*th clones of the two original federates, respectively. The federates are tailored based on the DMSO standard benchmarking programs [82].

As indicated in Figure 7.5, each pair of *Fed A[i]* and *B[i]* comprises an exclusive federation, which is denoted as *RTI[i]*. We employ this set of scenarios to perform the benchmarking experiments for the MF solution. In Figure 7.6, all federates form one single federation; accordingly, we use this set of scenarios to measure the performance of NDSF and DSF solutions.

Each federate is time constrained and time regulating, or neither. In one run, each federate updates an attribute instance and receives an acknowledgment from its partner (from *Fed A[i]* to *Fed B[i]*, and vice versa) for 10,000 times with a payload of 150 bytes. A federate merely reflects the events with identical ID to itself. In other words, *Fed A[i]* will discard any events not generated by *Fed B[i]*, and vice versa. In TSO mode, federates advance federate time from 0 to 10,000 with timestep = 1 and lookahead = 1. To investigate the efficiency for time synchronization among federates, we also examine the execution time for the standard Time Advancement benchmarking federates using both the MF and SF solutions.

Each federate can also be set as DDM enabled and Non-DDM. An exclusive ID is shared between *Fed A[i]* and *Fed B[i]*. In DDM-enabled mode, each pair of federates has an associated region, which is pair-specific and non-overlapping to any other region. As previously mentioned, the **StaticGrid-Partitioned** strategy partitions individual spaces into a grid in which each grid cell is assigned a separate reliable and best-effort channel [63]. The size of the grid is specified by a parameter *NumPartitionsPerDimension (NPPD)*. To optimize the utilization of communication channels, we split the full dimension $[MIN\_EXTENT, MAX\_EXTENT)$ evenly into the same number of segments as *NPPD*, which is set to the number of scenarios (federate pairs). The middle point of each segment will be defined as one region. We set the region associated to the kth pair of federates as

$$\frac{(2k-1)(MAX\_EXTENT - MIN\_EXTENT)}{2 \times NPPD} + MIN\_EXTENT,$$

$$\frac{(2k-1)(MAX\_EXTENT - MIN\_EXTENT)}{2 \times NPPD} + MIN\_EXTENT + 1)$$

In these experiments, all federates subscribe and publish the same object class, and associate the designated region to their subscription and updates.

7.4.2   Benchmark Results and Analysis

Some abbreviated notations are used to denote the properties of the federates, as listed in Table 7.2. In order to investigate the factors that impact the performance of alternative solutions, several series of experiments are designed as indexed in Table 7.3. The notations are the same as in the previous discussion. Combining the time features and solutions together, we perform eight series of experiments in total. In each series of experiments, the number of federates increases by one pair each time from five pairs to fourteen pairs.

TABLE 7.2
Notations of the Federate Attributes

Notations	Meaning
TSO	Federates are Time Regulating /Time Constrained and use Time Stamp Order update and reflection
RO	Federates use Receive Order update and reflection only
SYN	Standard DMSO time advancement benchmarking application
MF	Experiment in Multiple-Federation mode
DSF	Experiment in Single-Federation mode using DDM
NDSF	Experiment in Single-Federation mode without using DDM

TABLE 7.3

Index of the Experiments

	TSO	RO	SYN
NDSF	Experiment 1: TSO-NDSF	Experiment 4: RO-NDSF	Experiment 7:SYN-SF
DSF	Experiment 2: TSO-DSF	Experiment 5: RO-DSF	
MF	Experiment 3: TSO-MF	Experiment 6: RO-MF	Experiment 8: SYN-MF

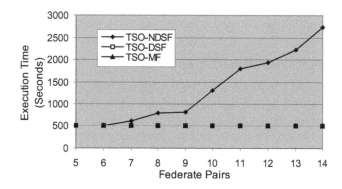

FIGURE 7.7

Execution time comparison between different cloning solutions using TSO federates.

7.4.3 Comparing Alternative Cloning Solutions Using TSO Federates

Figure 7.7 reports the execution time of TSO federates using the three different solutions. The execution time of the TSO-NDSF scenarios has an obvious increase when the number of federate pairs reaches seven. When no DDM services are used, the execution time increases sharply with more and more TSO federates joining the same federation. The topmost value (2,800 seconds, 14 pairs) is over 5 times the start value (~500 seconds, 5 pairs). On the contrary, the execution times of both the TSO-DSF and TSO-MF scenarios stay at a relatively stable level, about 500 seconds, in spite of the increase in the number of participating federates.

In Single-Federation mode, all federates interact with each other through

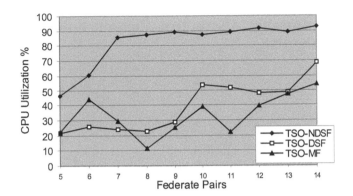

FIGURE 7.8
CPU utilization comparison between different cloning solutions using TSO federates.

the same RTI, the computation load of LBTS increases with the number of federates. In the TSO-NDSF scenarios, of more importance is that each federate not only receives useful events from its partner federate but also has to filter out some useless messages from all other federates as they belong to other scenarios. The overall communication traffic through the RTI is proportional to $C_n^2 = \frac{n(n-1)}{2}$, where n is the number of federates. Only $\frac{1}{(n-1)}$ of the total incoming events make sense to one particular federate pair. Other unnecessary communication and reflection increase the overhead in the RTI dramatically. The DDM services strictly confine the interaction to the pair of federates with a common region. This optimization results in the significantly improved performance as indicated in the curve "TSO-DSF."

In Multiple-Federation mode, a federate interacts with its partner through an exclusive RTI. Also, the LBTS computations will take place between a pair of federates independently. These positive factors lead to the much better performance compared with the TSO-NDSF scenarios.

CPU utilization percentage reports the processor activity in the computer. This counter sums the average non-idle time of all processors during the sample interval and divides it by the number of processors. The CPU utilization results in Figure 7.8 indicate that the TSO-NDSF scenarios consume much more system resource than the other two solutions.However, the TSO-DSF and TSO-MF scenarios also have an uptrend in terms of CPU utilization. The CPU utilization of the TSO-NDSF scenarios reaches about 90% after the number of federate pairs exceeds seven. These experiments are a combination of complex executions, including LBTS and TSO events receiving and reflecting. We attempt to isolate these two factors in the following experiments.

7.4.4   Comparing Alternative Cloning Solutions Using RO Federates

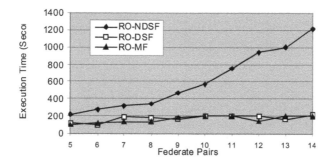

FIGURE 7.9
Execution time comparison between different cloning solutions using RO federates.

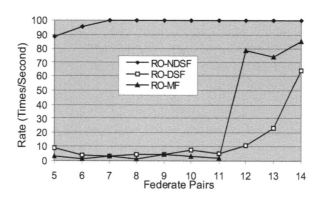

FIGURE 7.10
CPU utilization comparison between different cloning solutions using RO federates.

In order to further investigate the computational Complexity in these solutions, we disable the time feature of federates and reapply the three solutions to them. Execution time and CPU utilization are presented in Figure 7.9 and Figure 7.10, respectively.

Similar to the previous experiments, the execution time of RO-NDSF scenarios increases rapidly with the number of federates. The peak execution time ($\sim$1200 seconds, 14 pairs) is about six times the start value ($\sim$200 seconds, 5 pairs). The execution time of RO-NDSF scenarios always has a greater value than that of RO-DSF and RO-MF scenarios. The RO-DSF and RO-MF scenarios have execution times that fluctuate slightly from 100 seconds to 200 seconds. From the discussion of the TSO scenarios, the extra communication and reflection lower the performance of RO-NDSF solutions significantly.

Figure 7.10 also shows that the RO-NDSF scenarios consume much more system resource than the other two solutions. The CPU utilization of the RO-NDSF scenarios jumps to 100% after the number of federate pairs exceeds six. The RO-DSF and RO-MF scenarios have a very low CPU utilization (less than 10%) until there are more than twelve pairs of federates.

7.4.5   Comparing Alternative Cloning Solutions Using Time Advancement Benchmark Federates

In order to have a better understanding of how another factor, the LBTS calculation, impacts the performance, we reexamine the SF and MF solutions by introducing the standard Time Advancement Benchmark Federates [82]. Execution time and CPU utilizations of both solutions are shown in Figure 7.11 and Figure 7.12, respectively. As DDM does not intervene in the synchronization among federates, this series of benchmarks ignores the DDM mechanism.

We can conclude that the LBTS calculation does not make a significant difference in the execution time of either the MF or SF scenarios with an increase in the number of federates. The RO-NDSF and TSO-NDSF scenarios show a fast increasing execution time. This means that the reduction in interaction is the key to the optimization of the overall system performance in the simulation as long as the number of federates stays in a reasonable range.

7.5   Summary

This chapter proposed alternative solutions for distributed simulation cloning, namely the Single-Federation solution (with and without DDM) and the Multiple-Federation solution. The alternative candidate solutions were compared in terms of robustness, Complexity, and efficiency.

In the context of the Multiple-Federation (MF) solution, multiple federations operate in parallel to explore alternative scenarios. Thus, the MF solu-

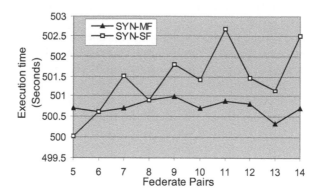

FIGURE 7.11

Execution time comparison between cloning solutions using time advancement federates.

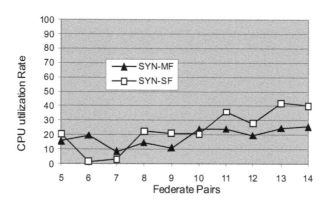

FIGURE 7.12

CPU utilization comparison between cloning solutions using time advancement federates.

tion surpasses in robustness and federate synchronization. If a failure occurs in a scenario, the failure will not affect the remaining scenarios as they are in different federations. However, scenario management becomes more complex. In particular, the requirement of dynamic clone sharing by incremental cloning is extremely difficult in the MF solution as the shared clones need to operate in different federations. The Single-Federation solution (SF) thus has an advantage in cloning control and clone sharing.

The benchmark results indicate that, in general, MF exhibits much better performance than NDSF. The interaction among federates is the main factor that impacts the execution speed. In the case where there is a high data exchange, the performance can be improved dramatically by applying DDM to NDSF. The calculation of LBTS makes no significant difference in either the MF or SF solution. The results show that DSF performs as well as MF in terms of efficiency. Considering the reduction in implementation Complexity and convenience for scenario management, the Single-Federation solution applying DDM (DSF) was chosen for cloning distributed simulations in this current study.

8

Managing Scenarios

CONTENTS

A federate spawns multiple clones to explore different new scenarios at Decision points; thus the overall distributed simulation comprises multiple concurrent scenarios.Data Distribution Management (DDM) provides a method to partition concurrent scenarios in an HLA-based distributed simulation. This chapter details our design of two DDM-based solutions and analyzes their advantages and drawbacks.

8.1   Problem Statement

As multiple dynamic scenarios operate in a common overall distributed simulation, a mechanism is necessary to identify the scenarios and partition them, so as to manage them in a precise and efficient manner. Moreover, to minimize the overhead incurred by creating more and more scenarios, we need to reduce bandwidth requirements by only transmitting interactions or attribute updates where necessary. As a shared clone clone may operate in different scenarios dynamically, as described in Chapter 6, there needs to be a dynamic event "routing" and "filtering" mechanism to support cloning. To provide reusability of existing simulations, it is also necessary to hide the complexity of the above operations. In an HLA-based simulation, this means we need to maintain the standard HLA interface to the user's programs while providing extended functionalities.

This chapter extends the idea of using the Data Distribution Management (DDM) mechanism to partition concurrent scenarios in a distributed simulation. In the benchmark experiments on DDM and Non-DDM cloning solutions in Chapter 7, we compared the performance of these two approaches in terms of execution time using both Time Stamp Order and Receive Order cases. The performance of the DDM enabled approach significantly exceeds the Non-DDM one. This superiority is obvious in a large-scale distributed simulation.

Two alternative solutions are introduced in this chapter for managing scenarios and identifying each clone and scenario. A **recursive region division solution** and a **point region solution** are discussed and compared. The former solution initializes each original federate with a region occupying the full dimension of the routing space. During the cloning procedure, the new clones inherit split sub-regions from their parent. The latter solution specifies a point region for each original federate, and a new clone is given an additional point region on birth. Dynamic region combining is used for shared clones shared clones. Our potential solutions for managing scenarios aim to provide the standard interface of Object Management (OM) or DDM services while employing additional DDM methods in the underlying middleware. Thus, it is possible to hide the DDM solution implementation behind the normal OM or DDM services interface in a transparent way. The solutions to be discussed focus on addressing the manipulation of regions and the mapping between each region and the scenarios.

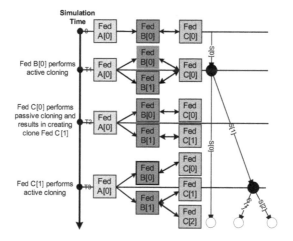

FIGURE 8.1
Formation of a scenario tree.

As discussed in chapter 6, the active cloning of federates incurs the spawning of new scenarios in the simulation. The concurrent scenarios can be represented as a tree developing along the simulation time axis. Figure 8.1, which is a simplified version of Figure 3.4, illustrates how a scenario tree is formed. The issues to be considered in comparing the two alternative solutions include coding scenarios, region specification for each scenario, and implementing the middleware approach. To illustrate the details of each solution, the example in Figure 8.2 is also used in the following sections for studying the coding scheme. Seven scenarios (labeled "a" to "g") are present in the overall simulation following two active clonings at times T_1 and T_2, respectively.

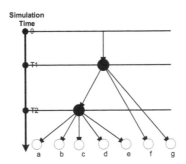

FIGURE 8.2
An example of a scenario tree.

In order to minimize the computation involved in DDM, we use only one single routing space having a single dimension for cloning. To ease discussion, we assume the federates being studied do not use DDM services in their models. However, federates that already use DDM services can also easily apply the solutions without changing the federate code. This can be achieved by associating another "cloning" dimension in the middleware to the existing DDM-enabled interactions and attribute updates when necessary.

The underlying middleware subscribes and publishes with the same region for a given clone. Each scenario is associated with an exclusive region, and there is no overlap with any other scenario's region at any point during execution. Clones within the same scenario utilize a common scenario-specific region extent[1] unless they are shared clones. A shared clone has a merged region that exactly covers the scenario-specific extents of all the scenarios in which it operates. DDM also aids interactive control of the scenarios at runtime; external commands can be easily routed to the clones within a given scenario by associating the proper region to them. The approach will lessen the work of recognition and processing at the clone side.

---

[1]In this book, extent is used to mean the interval [Lower_Bound, Upper_Bound] that defines the region.

8.2   Recursive Region Division Solution

The basic idea of the recursive region division solution is to divide the full dimension (i.e., [MIN_EXTENT, MAX_EXTENT]) from top to bottom. A federation is initialized with all federates having an associated region with extent [MIN_EXTENT, MAX_EXTENT]; thereby the initial scenario has a full dimension region. Once new scenarios are created, each scenario inherits a sub-region from its "parent" scenario under a region division algorithm. Thus the active cloning federate should divide its original region for the child clones, while its partner may still keep the region unchanged. In the example shown in Figure 8.1, immediately after time *T1*, *Fed B[0]*'s original region is split into two parts; *Fed B[0]* and *Fed B[1]* modify their region with the first and the second part, respectively. The new regions indicate that *Fed B[0]* and *Fed B[1]* belong to scenario *S[0]* and *S[1]*, respectively; however, *Fed C[0]* will remain associated with the region of both scenarios *S[0]* and *S[1]*. Thus the events from both *Fed B[0]* and *Fed B[1]* will be automatically routed to *Fed C[0]* without any modification to its region. This solution requires minimal work for shared clones. Region division keeps taking place along the development of the scenario tree.

The scenario tree can be reconstructed as a binary tree to ease coding. When *n* new scenarios are developed from one scenario branch, the following rule will be applied:

Given that $2^{k-1} < n \leqslant 2^k$, *n* is rewritten as $n = 2^{k-1} + m, m \leqslant 2^{k-1}$. Then the branch extends *k* levels downward, the leftmost *m* nodes at level $k - 1$ will always be expanded one further level.

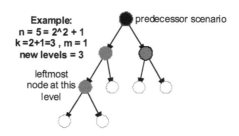

Example:
n = 5 = 2^2 + 1
k = 2+1 = 3 , m = 1
new levels = 3

leftmost node at this level

predecessor scenario

FIGURE 8.3
Development of binary scenario tree.

Following this rule, the scenario tree in Figure 8.2 is converted to a binary tree. Suppose that five new scenarios are created from the predecessor scenario node at time *T2*; the development of this branch is as indicated in Figure 8.3. The leaf nodes represent the newborn scenarios including the original one.

As shown in Figure 8.4(A), we code a node's left branch as "0" and the

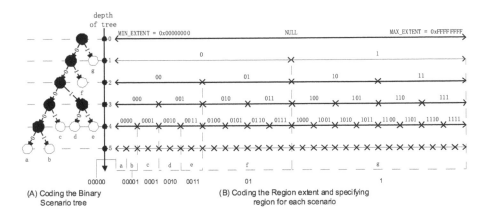

FIGURE 8.4

Binary scenario tree and scenario region code.

right branch as "1" accordingly. Based on the binary tree, we link the branch codes together along the path from the root to the given leaf node; thus the exclusive code of each scenario is obtained. For example, scenario e is identified as "0011". The position of the scenario can be easily traced according to its identity.

A shared clone needs to distinguish events from different clones that are spawned by the same federate. We may also need to control the cloning procedure or update the system state of one particular clone. Therefore it is necessary to identify each clone accurately. A clone always joins one or multiple scenarios; naturally the clone identity needs to cover this information. A direct scheme is to combine scenario codes and the federate name of the original ancestor to this clone, with format "<Scenario ID>&&<Federate Name>". For example, a clone joins scenario e and its ancestor federate is named "fab1"; then this clone may be coded as "0011&&fab1". In the case of shared clones, a delimiter symbol is placed between scenarios; if necessary, a wildcard is also introduced to reduce the identity length. The format is "<Scenario ID 1>:<Scenario ID 2>: ... :<Scenario ID N>&&<Federate Name>", in which ":" is a delimiter and "*" is a wildcard. For example, if another clone of ancestor "fab2" is shared among scenarios a, b, and g, this clone is coded as "0000*:1&&fab2".

Figure 8.4(B) gives the region-coding scheme. The relationship between the position of a scenario node and its specific region is illustrated explicitly. The full dimension is segmented more and more densely with the increase in level depth.

At the kth level (mapping the nodes in the binary tree with depth equal to k), the full dimension is evenly partitioned into $n = 2^k$ segments. Each segment is given a binary code according to its index, with the code length equal to the level depth. The segment code is length sensitive; for example, the code "011" at level 3 differs from the code "0011" at level 4. For the purpose of illustration, assume that MIN_EXTENT = 0x00000000 and MAX_EXTENT = 0xFFFFFFFF; then the following simple formula gives the calculation of the extent of the ith segment at level k:

$$\text{LowerBound} = i \times 2^{32-k}$$
$$\text{UpperBound} = (i+1) \times 2^{32-k} - 1 \tag{8.1}$$

The code of each scenario coincides perfectly with the segment extent . The extent of the scenario-specific region can be directly obtained from the scenario code. First we assign the length of the scenario code to k; second, we calculate $i= atol(scenario\ code)$; then the extent of the region is available immediately from Equation (8.1). For example, the code of scenario e is "0011"; then we have $k = strlen("0011") = 4$ and $i = atol("0011") = 3$:

$$\text{Lower Bound} = 3 \times 2^{32-4}$$
$$\text{Upper Bound} = (3+1) \times 2^{32-4} - 1$$

The region extent of scenario e is [0x30000000, 0x3FFFFFFF]. The one-to-one map between a scenario code and the region extent is constant for any simulation. This feature implies that we do not have to record the map between a scenario ID and its specific region extent. Using the recursive region division solution avoids the need for a searching procedure.

The full dimension [MIN_EXTENT, MAX_EXTENT] contains 2^{32} unique extents at most. Thereby, 2^{32} concurrent scenarios are allowed when making full use of the dimension. Such a large number is able to meet any practical requirement for classifying scenarios. However, the binary division algorithm may incur problems in some extreme situations. Let us look at a special example in which on active cloning only the leftmost branch in the binary scenario tree splits into multiple new scenarios. Thus the region extent of the leftmost scenario will shrink rapidly in an exponential way, much faster than other scenarios. It can be seen that region allocation is densely concentrated at the left end of the dimension, whereas only a few scenarios occupy the remainder of the dimension. As this continues, when the depth of the binary tree reaches 32, one scenario's region extent becomes a point and child scenarios are not able to inherit extents any more. This potential limitation exists even if this is unlikely to occur.

Once the extent is exhausted in the single clone dimension, one possible solution is to redistribute the extents of the full dimension. Undoubtedly the redistribution incurs extra complexity and it damages the natural harmony of the one-to-one map between scenario identity and its region extent. Alternatively we can specify a multidimensional routing space for cloning beforehand. A clone region is created with these dimensions, and the recursive division algorithm is initially applied to the first dimension. When the extent is exhausted in this dimension, the subsequent dimensions are still available for extent allocation using the same recursive division algorithm. This approach is able to keep the coding scheme intact. As long as the number of clone dimensions is set large enough, it should meet any request in practice.

8.3 Point Region Solution

An alternative solution, namely the point region solution, is proposed to distribute the extents of the dimension as evenly as possible in a bottom-up manner. A point region has an extent defined as [Lower_Bound, Lower_Bound+1). As the name implies, the point region is a single element used in associating a region with a scenario. The initial federates are assigned a start point region. Instead of inheriting any region from their predecessor, the new scenarios get different point regions from the dimension. A shared clone has to combine its original region with the new ones associated with additional scenarios.

Thus, scenarios can be coded based on the scenario tree. Figure 8.5 illustrates a coded scenario tree based on the example given in Figure 8.2. In this solution, on an active cloning, no matter how many new scenarios are created from an existing scenario, the scenario tree extends only one level down with all the new scenarios (and the existing one). From left to right, sibling branches of any scenario are labeled with "0" to "n". For the scenarios that remain unsplit, in the scenario tree, the corresponding nodes still extend one level down with one single child to keep consistency in representing current scenarios and the cloning history along the simulation time axis. By simply linking codes starting from the root, the identity of a scenario is obtained. To ease representing and resolving the scenario identity, we can record the label of each branch in hex format, where each byte holds a sibling's code; thus scenario e's identity is written as "0004". The clone identity follows the format defined in the previous recursive division solution. Given the same examples as in Section 8.2, we can code the descendant clone of original federate "fab1" that joins scenario e as "0004&&fab1". As for another clone with ancestor "fab2", which is shared by scenarios a, b, and g, we can code it as "0000:0001:0200&&fab2". This shared clone has a region that is the union of region extents associated with scenarios a, b, and g.

To optimize the usage of Communication channels provided by DDM in RTI-NG, the single "cloning" dimension is evenly divided into multiple segments according to the *NumPartitionsPerDimension (NPPD)* in the "RTI.rid" file [1]. Each segment will contain an identical number of point regions. To ease discussion, here we still assume that MIN_EXTENT = 0x00000000 and MAX_EXTENT = 0xFFFFFFFF. The following matrix describes the distribution of all available point regions with the number of rows equal to *NPPD* and the number of columns equal to (MAX_EXTENT-

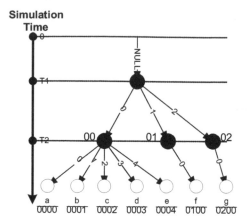

FIGURE 8.5
Coding the scenario tree.

MIN_EXTENT+1)/$NPPD$. It is obvious that the points in one row belong to the same segment in the dimension.

$$
\text{Rgn} = \begin{bmatrix} R_{0,0} & R_{0,1} & \cdots & R_{0,(2^{32}/NPPD-1)} \\ R_{1,0} & \cdots & & \vdots \\ \vdots & & & \vdots \\ R_{(NPPD-1),0} & \cdots & & R_{(NPPD-1),(2^{32}/NPPD-1)} \end{bmatrix} \tag{8.2}
$$
$$
where \quad R_{i,j} = (2^{32}/NPPD) \times i + j
$$

In total, there can be up to 2^{32} available point regions in which the jth point region in the ith row is written as $[R_{0,0}, R_{0,0}+1)$. The initial scenario starts with a region $[R_{i,j}, R_{i,j}+1)$. A new scenario will always be assigned the first unused point region in the next row. Once the regions in the same column are fully used, the allocation will start from the beginning of the next column. Figure 8.6 illustrates this region specifying flow as indicated by the arrowed line. The scenario ID maps to its specified region in a one-to-one way. The scenario tree records the scenario ID and the region in the node for that scenario, and can be used for resolving the region of one particular scenario from its identity by searching the tree. The scenario ID is resolved to locate the scenario node in the tree; this search is only performed along a given path instead of making a full search of the scenario tree.

The point region solution has an added advantage in that it maximizes the use of available communication channels. For each routing space, RTI-NG will create reliable $NPPD^{(Number\,of\,Dimension)}$ and best-effort channels. The channels are mapped to the dimension in a grid-like fashion. As we only define one dimension, RTI provides exactly $NPPD$ pairs of channels for the

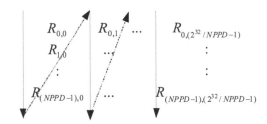

FIGURE 8.6
Point region allocation flow.

"cloning" space in our solution. The point regions in the same row (see the matrix in Equation (8.2)) occupy a common data channel. The point region ensures that the interactions or updates with which it is associated use a unique channel when possible and avoids the overhead of sending data through multiple channels unnecessarily. However, this solution does not support a direct conversion between the scenario ID and region; the scenario manager module needs to record this map relationship. Furthermore, shared clones need to modify their regions on the latest active cloning. This incurs extra effort compared with the recursive region division solution.

8.4 Summary

To address the complexity of the overall cloning-enabled distributed simulation due to increasing scenario spawning, we investigated an efficient and precise scheme to identify and represent scenarios. This chapter described and compared two alterative scenario management algorithms, namely the recursive region division solution and the point region solution. The two solutions were designed to code scenarios as well as to manipulate region extents in the "cloning" dimension. The former solution initializes each original federate with a region occupying the full dimension of the routing space. During the cloning procedure, the new clones inherit split sub-regions from their parent. The latter solution specifies a point region for each original federate and a new clone is given an additional point region on birth. Shared clones may have changing region combinations when cloning occurs in the scenarios in which they operate.

Table 8.1 gives a concise comparison between the recursive region division solution and the point region solution. The recursive region division solution has some advantages in that (1) it avoids the extra operation of manipulating regions for shared clones which is required by the point region solution and

TABLE 8.1

Comparison between the Recursive Region Division and the Point Region
Solutions

Features	Recursive region division	Point region
Characteristic region	An extent	A point
Region Manipulation for cloning	Top-down division, a clone inherits region from its parent	Bottom-up distribution, each clone's region is independent
Number of scenarios that can be handled	In the worst case, 33 times the number of "cloning" dimensions	2^{32}
Support of clone sharing	Clone sharing is facilitated automatically	Extra region manipulation is required for shared clones
Map scenario ID to region	Natural one-to-one map	Specific map is needed
Use of communication channels	Not optimized	Optimized

(2) it provides a natural one-to-one mapping between scenario identity and region extent. The coding mechanism harmonizes with the region specification perfectly. However, the recursive region division solution has a limitation in dealing with some extreme situations. The point region solution has advantages over the former one in that (1) it can meet the region allocation requirements even in extreme situations; and (2) it optimizes the use of data channels. The latter solution is therefore chosen and developed in designing our cloning technology due to these superior features.

9

Algorithms for Distributed Simulation Cloning

CONTENTS

The cloning management mechanism is designed to ensure correct federate replication, and its tasks include creating clones, manipulating states, and coordinating the federation. This chapter first introduces the overall cloning infrastructure, and details the algorithms for distributed simulation cloning, including active cloning and passive cloning. Subsequently, this chapter discusses incremental cloning and the algorithms for managing shared clones.

9.1   Overview of Simulation Cloning Infrastructure

An RTI++ infrastructure enabling simulation cloning is built as the middleware[1] between the simulation model and the real RTI. The infrastructure contains several major modules that perform the necessary functionalities related to simulation cloning while presenting a standard RTI interface to the simulation model (as shown in Figure 9.1).

Prior to using simulation cloning, the user can specify the conditions

---

[1]The physical federate can also be regarded as a component of the whole middleware between the simulation model and the real RTI. In this discussion, we are only concerned with the RTI++ library that directly interacts with the simulation model in the virtual federate.

and/or rules according to which the cloning should be triggered and the different candidate actions to be taken.

The Control Module monitors the states in which the user is interested and evaluates the conditions for cloning the federate at a decision point. The Cloning Manager module creates new clones for the request issued by the Control Module, and it initiates the creation and update of the scenarios. The Scenario Manager module creates and stores the scenario tree, from which the identity and corresponding DDM region of each scenario can be fetched. The Scenario Manager keeps the history of the overall cloning procedure. The Region Manager module creates DDM regions and manages the regions. The Region Manager services the Scenario Manager by answering enquiries and providing the clone specific region. The Region Manager also deals with any request for modifying a region. The Scenario Manager presents the region to the middleware for partitioning event transmission. The RTI++ Object Management services invoked by a federate are executed via the corresponding DDM methods by associating the region obtained from the Scenario Manager. Eventually it is the physical federate who calls the real RTI services and conveys the callbacks to the Callback Processor in the RTI++ middleware (see Chapter 3).

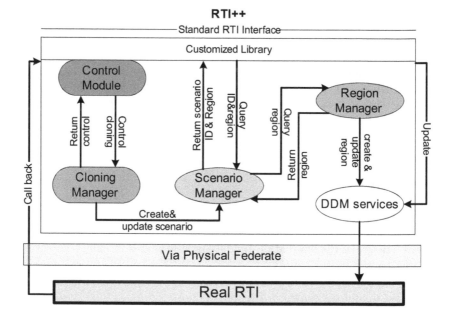

FIGURE 9.1
RTI++ and internal modules.

In the context of the middleware approach, the Cloning Manager provides services directly to the Control Module that handles the decision point and cloning trigger. It also interacts with the Scenario Manager (see Chapter 8) to update/fetch scenario information, such as region and scenario ID etc. Once the condition for cloning is met, the cloning trigger invokes the Cloning Manager and the latter performs cloning of the federate accordingly. External stable storage is used to store RTI States of the federate. Figure 9.2 illustrates the internal components inside the Cloning Manager, which is designed to

- Log and update the system state at the RTI level to stable storage.

- Replicate the virtual federate process at runtime according to the parameters given by the cloning trigger.

- Coordinate with other federates for the cloning operation.

- Recover system states to initialize the physical federates of all clones.

- Link the cloned virtual federates to the corresponding physical federates and resume simulation execution.

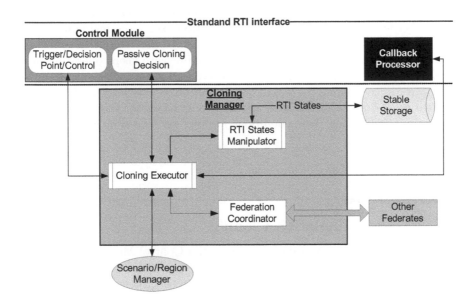

FIGURE 9.2
Cloning Manager module.

The Cloning Executor works as the key component of the Cloning Manager module. It answers the cloning request issued by the Control Module. Upon cloning, it interacts with the Scenario Manager to update and retrieve scenario and characteristic region information. The Cloning Executor makes replicas of the simulation model. In addition, it initiates the same number of new physical federate instances. The saved RTI States are loaded from the external stable storage to initialize the physical federates. Once the state of each clone is initialized to the snapshot of the parent federate at the decision point, the Cloning Executor resumes the whole simulation after applying the action defined for the decision point to each clone. The Cloning Executor also handles the in-transit events on cloning using the RTI federation save services, which forces all in-transit events to be delivered to all receivers.

The RTI States Manipulator saves RTI states and replicates them when needed. The RTI states include federation information, publication/subscription information, object registration information, time granted, etc. A new clone needs to register object instances that represent the same entities known to the simulation model in the parent federate. However, the RTI assigns different handles [72] to the "cloned" objects, which are unknown to the simulation model. The RTI States Manipulator deals with this problem using an entity mapping approach. The approach maintains the relationships among original entities and "cloned" entities at the RTI level. All incoming events will be mapped (processed) by the Callback Processor (see Chapter 3) prior to conveying them to the simulation model.

During cloning of a federate, the Federation Coordinator should synchronize other federates within the whole simulation execution, including the sibling clones. After cloning is initiated, execution of the simulation model of a federate (clone) will be paused, while the middleware keeps operating in coordination with the cloning procedure. The coordinator is designed to ensure that prior to continuing the simulation execution, every clone has finished initialization and all federates have updated their states for the current cloning operation. Only when these conditions are met can the simulation be reactivated and resumed.

9.2   Active Simulation Cloning

When a federate reaches a decision point, the Control Module evaluates the states in which the user is interested. Once the cloning conditions are met, the Cloning Manager will make clones immediately to explore alternative executions as predefined in the cloning triggers. Then an active simulation cloning occurs. The partner federates may perform passive cloning or operate as shared clones.

From the perspective of the federate (parent federate) making clones, the process of an active simulation cloning can be described as follows (see Figure 9.3). In this process, the RTI services used by the middleware are directly executed by the physical federate, remaining transparent to the simulation model.

- **Initiating cloning**. The Cloning Executor calls *cloningManager::createClone (number of clones)* to initiate the active cloning. The federate being

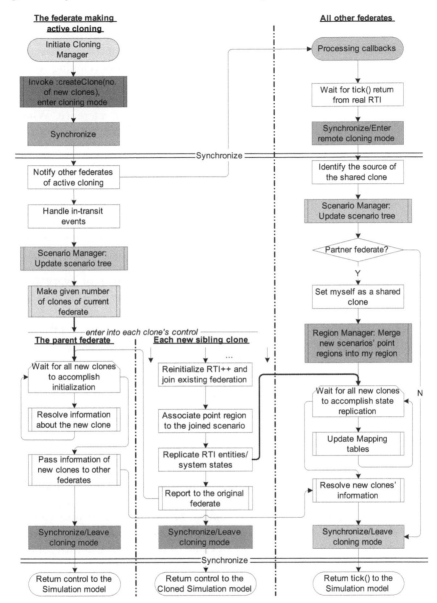

FIGURE 9.3
Active simulation cloning.

replicated enters Cloning mode, in which control is with the RTI++ middleware.

• **Synchronizing federation**. The Federation Coordinator notifies and synchronizes the remaining federates by requesting a federation save. The federation save label is coded to contain the following information: (1) whether the current simulation cloning is an active one or passive one, (2) the number of new clones to be created, (3) the handle of the federate making clones, and (4) the scenario within which cloning occurs (cloning scenario). The remaining federates retrieve this label via their callbacks. When they have identified that this synchronization is for the purpose of cloning, their Callback Processors extract the coded information for reference. When the whole federation has completed synchronization, all federates enter cloning mode.

- **Handling in-transit events**. The current design utilizes the federation save services to ensure all in-transit events reach their destinations prior to cloning. The Callback Processor will present those events to the virtual federate when cloning is accomplished. Those events with timestamp greater than the current federate time must be passed to the simulation model with the advancing of simulation time.

- **Updating scenario tree**. The Scenario Manager of every federate updates its scenario tree (see Chapter 8) based on the cloning scenario ID and clone number. The scenario updating algorithm ensures all federates maintain identical scenario trees. Those federates remaining intact check whether they operate in the existing scenario as a partner federate or not. A partner federate will work as a shared clone in the current and new scenarios; the region manager needs to merge the new scenarios' point region extents into its existing region and notify the RTI about this region update. The updated region can be written as $region(Scen[1]) \cup \ldots \cup region(Scen[n])$, in which $region(Scen[i])$ represents the characteristic point region associated with the ith scenario in which the shared clone will execute. Non-partner federates directly block until the end of the current cloning.

- **Making clones**. The Cloning Executor makes the specified number of replicas of the simulation model and initiates an individual physical federate for each replica. Within each replica, the Cloning Executor causes the virtual federate to link to the corresponding physical federate, which joins the existing federation. Thereafter, a clone of the parent federate comes into being, which leads to the construction of a new scenario.

- **Replicating system states**. A clone's physical federate must be initialized with the stored system states from the parent federate. The system states to be replicated include (1) object classes/interactions that have been published/subscribed, (2) registered object instances, and (3) federate time. The state replication of the new clones is detailed later. The parent federate and partner federates block for the new clones to complete state replication. Each clone reports to the parent federate when fully initialized with its inherited system states. All clones' identities will be collected and forwarded to the partner federates by the parent federate.

• **Leaving cloning mode**. After all clones announce that they are ready, the Federation Coordinator of the parent federate initiates another federation synchronization to declare the termination of the cloning process. The synchronization label codes the relationship between each clone's federate handle and the scenario in which it will operate. The partner federates' Callback Processors can decode the mapping relationships and use them to distinguish the source of incoming events in the future. As soon as synchronization is achieved, the parent federate, its clones, and the remaining federates obtain control from the middleware.

After leaving cloning mode, the paused simulation execution resumes with the new scenarios starting operation. The new clones and the parent federate each execute in a single scenario, while the partner federates of the parent initially become shared clones that execute in the original scenario and the new ones. All federates work in normal mode until the next occurrence of simulation cloning.

Figure 9.4 depicts the state replication procedure of a new clone. This procedure involves the parent federate, the new clones, and the partner federates, but does not affect the remaining federates. Prior to creating clones, the parent federate sends a "system" interaction (namely, "*cloningNotification*," defined for cloning) to the partner federates. This interaction encapsulates the published object/interaction classes and the list of registered objects of the parent federate. The receiver's Callback Processor can simply take the intersection of this remote object list and the original object instances discovered previously, and then it can predict what image object instances (see Section 6.3.1) it should discover from each new clone. Hence, information contained in this interaction provides criteria for the partner federate to detect the state replication status of each clone. Furthermore, the receiver's Callback Processor will refer to this remote publication and registration information in processing events from the parent federate's sibling clones.

A new clone's state replication can be broken down into three phases. When a clone is created, first it enables time constrained and/or time regulating, depending on the parent federate's settings. The federate time and lookahead recorded on cloning should be specified accordingly.

Second, the clone's Cloning Manager associates a scenario-specific point region to the clone, which is obtained from the scenario tree according to the scenario in which the clone will execute. The States Manipulator instructs the physical federate to subscribe/publish the same object/interaction classes as the parent federate has subscribed/published, but with the newly retrieved region. The partner federates (shared clones) merge the designated point regions of the new scenarios into their current region.

Third, the **new clone** needs to register image objects to the objects owned by the parent federate. The physical federate invokes the RTI method *RTI::registerObjectClassInstanceWithRegion(objectClass, Name, ...)* [72] and conveys the returned object handle to the middleware, and the latter maps the original objects to the image ones. The specified "name" of an image

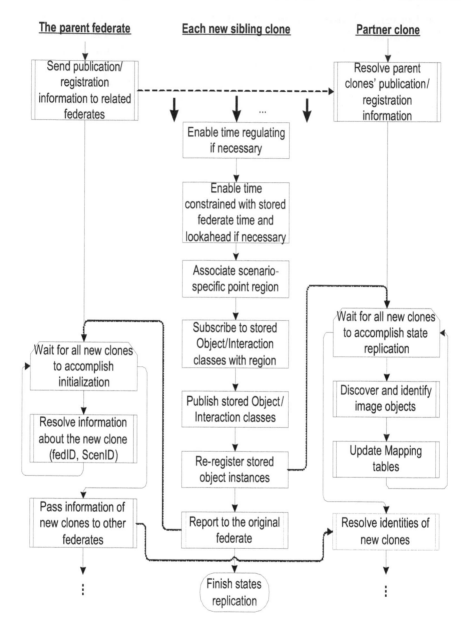

FIGURE 9.4
Replicating states for new clones.

object follows the format "<federate handle>&&<original object instance handle>". The middleware of other **partner federates** that are subscribers will discover these image objects (their regions overlap with the clone's region)

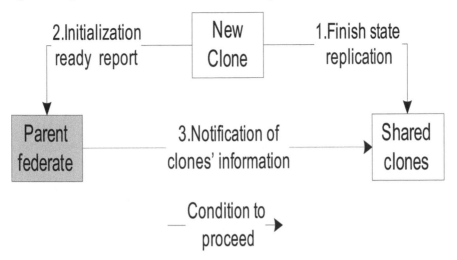

FIGURE 9.5
Coordinating federates in state replication.

and can map them to the original ones from their names. The mapping relationship (recorded in Mapping Tables) will be used by the Callback Processors to process events from different scenarios. The Callback Processor blocks until all image objects have been discovered, and hides these object discoveries from the simulation model to keep correct HLA semantics. Otherwise, the simulation model will "rediscover" object instances it has already discovered before cloning. On the other hand, the **new clones** may also detect objects registered by the partner federates before cloning (due to the overlapped region). As the cloned simulation model has "already" discovered those objects, these object discoveries should also be hidden to avoid repeated discovery. The parent federate waits for new clones to be created and initialized. As soon as each clone completes initialization, it reports to the **parent federate** with a "*cloningReady*" message containing its federate handle and scenario ID. This design puts constraints (see Figure 9.5) on the parent federate and partner federates before leaving cloning mode. Only when all clones finish replicating states can the whole federation resume normal execution. Thus the state consistency of the overall simulation can be achieved.

Let n denote the number of new scenarios to be explored at a decision point and r the number of RTI object instances to be re-registered. The computational complexity of the cloning process for an individual clone is $O(r)$. As the new sibling clones execute on the same processor as the parent federate, the overall computational complexity of the active cloning algorithm is $O(n \times r)$. This does not include the cost of communication and synchronization with other federates, which will depend on the underlying RTI implementation.

The overall performance of the cloning mechanism is analyzed experimentally in Chapter 10.

9.3   Passive Simulation Cloning

The partner federates of the federate making active cloning may remain intact and become shared clones in related scenarios, given that their system states are not affected by the cloning immediately. Incoming events of a shared clone are checked by the Callback Processor. The result of checking an event decides whether the shared clone remains shared or requires passive cloning. The event checking algorithm determines which events and how these events should be conveyed to the simulation model as well as when the passive cloning should be triggered.

As mentioned previously, a shared clone performs passive cloning to handle different events from existing scenarios rather than to explore new scenarios on its own initiative. The passive cloning does not cause creation of new scenarios, and this is different from active cloning. A passive cloning procedure is illustrated in Figure 9.6 from the shared clone's point of view, and can be described as follows:

- **Initiating passive cloning**. The Cloning Executor calls *cloningManager ::splitClone* to initiate the passive cloning, and the shared clone enters cloning mode. The in-transit events are handled in the same way as for active cloning.

- **Synchronizing federation**. The Federation Coordinator notifies and synchronizes with the remaining federates by requesting a federation save. These federates retrieve the federation save tag and identify the current passive cloning from the extracted information. When the whole federation has completed synchronization, all federates enter cloning mode.

- **Creating clones and updating region**. The Cloning Executor creates new clones, and the number of clones is set, depending on the number of scenarios in which the parent federate operates beforehand. The original region of the parent federate (denoted as $region(Scen[1]) \cup \ldots \cup region(Scen[n])$, in which $region(Scen[i])$) should be split into n individual point regions. The Region Manager of the parent federate and each new clone associates an exclusive point region to the clone according to the individual scenario in which it may participate. The parent federate becomes a normal clone and so do its new clones, while the remaining federates continue to operate as normal or shared clones. Their existing scenarios are kept intact without generating new ones.

- **Replicating system states** (see Figure 9.4). The mapping tables of partner federates will be updated by their Callback Processors according to the re-registered image object instances.

• **Leaving cloning and conveying buffered callbacks**. After each clone finishes initialization, another federation synchronization will be initiated to denote the ending of passive cloning. The Callback Processor of the parent federate or each new clone needs to convey the buffered events (received in pending-passive-cloning mode, see Section 9.5.2) to the Federate Ambassador. The events are inherited from the parent federate (for new clones). Only those events with timestamp not greater than the current federate time must be conveyed. Subsequently, each Callback Processor returns control to the simulation model and the latter obtains control again. Thereafter, the whole federation resumes normal execution.

Let n denote the number of independent scenarios from which a shared federate (clone) has received different events and r the number of RTI object instances to be re-registered. The computational complexity of the cloning process for an individual clone is $O(r)$. As the new sibling clones execute on the same processor as the parent federate, the overall computational complexity of the passive cloning algorithm is $O(n \times r)$. Again, this does not include the cost of communication and synchronization with other federates.

9.4 Mapping Entities

A federate simulation model produces or consumes information in its object model via RTI services using handles assigned by the RTI. For example, when an object instance is registered, a federation-unique handle is returned to identify that object instance. This handle is used to represent an entity known to the model and other federates that have discovered this object instance.

A clone inherits identical states from the original federate, including the RTI entities known to the simulation model. In order to keep the state consistent and federate code transparent, our cloning approach needs to ensure that the clones of a federate use the same reference to the original entities at the RTI level as before cloning. The approach should correctly manage the events associated with these entities within the overall federation; for example, a shared clone may receive updates of different object instances even though they refer to the same object in the simulation model (see Figure 9.7).

The consistency can be achieved using a mapping approach in the middleware. The middleware maps the original handles with the image object handles to ensure user transparency and consistency. For one original object instance referred to by the simulation models of all clones, there can be different image object instances accessed by the physical federates. The middleware keeps transparency of object instances in the object model by translating the

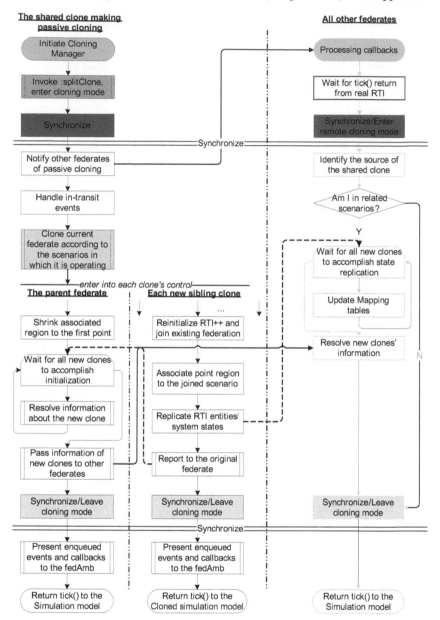

FIGURE 9.6
Passive simulation cloning.

handle as appropriate. The same principle is applied in processing other entities at the RTI level. An example of processing object instances is given in Figure 9.7. The physical federate is omitted in the figure to ease discussion.

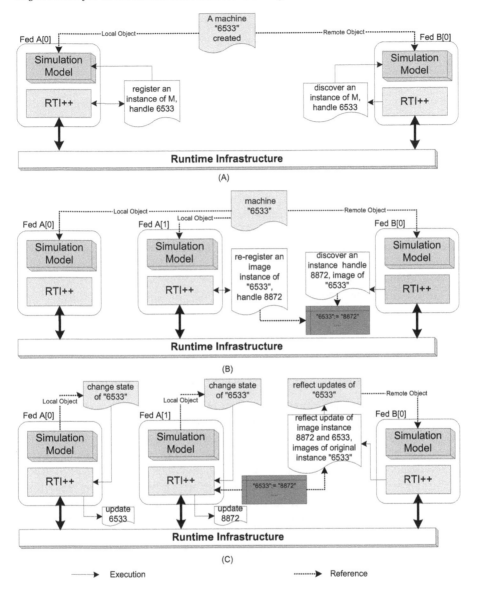

FIGURE 9.7
Mapping RTI entities.

The overall simulation starts with a federation formed by a pair of federates (*Fed A[0]* and *B[0]*). Assume that *Fed A[0]* publishes object class "machine" and *Fed B[0]* subscribes to it. As shown in Figure 9.7(A), when *Fed A[0]*registers an object instance of "machine" with object handle 6533 via its RTI++, its simulation model recognizes that a local machine object "6533" is created. *Fed B[0]* detects the object via its RTI++ and its simulation model also knows about the creation of this remote machine.

As illustrated in Figure 9.7(B), *Fed A[1]* is created as a clone of *Fed A[0]* at some point, and *Fed B[0]* is shared by *Fed A[0]* and *A[1]*. *Fed A[1]* re-registers an image object of class "machine" with the original object instance 6533 during state replication and gets a unique handle 8872 from the RTI. However, the simulation model of *Fed A[1]* still refers to the original object "6533", the original machine object. The object 8872 is an image object and is mapped to the original object 6533 by the RTI++. Similarly, *Fed B[0]*'s RTI++ discovers the new object and records its mapping to the original one. The new image object handle will not be passed to *Fed B[0]*'s simulation model, as *Fed B[0]* has already discovered the original object instance.

As simulation execution proceeds, *Fed A[1]* needs to change the state of machine "6533" (Figure 9.7(C)). Its RTI++ converts the update call of "6533" to the image object 8872 and passes it to the RTI services in the underlying layer. *Fed A[0]* invokes an update of image object 6533 directly on its simulation model's behalf. At the *Fed B[0]*'s side, these updates are automatically translated into the updates of machine "6533" for the simulation model.

9.5   Incremental Distributed Simulation Cloning

When a federate makes clones on its own initiative and creates new scenarios, other federates in the original scenario have to interact with each of these clones properly in the new scenarios. One direct solution is to clone all other federates immediately; thus, each independent set of clones forms a new stand-alone scenario. In this case, a full set of independent federates is exploited to examine each scenario after cloning. However, when performing distributed simulation cloning, it is desirable to replicate only those federates whose states will alter at a decision point. The remaining federates may keep intact and become shared between the original scenario and the new ones; only when absolutely necessary will those shared federates be cloned. Hence such an incremental simulation cloning mechanism is expected to further share computation among scenarios.

9.5.1   Illustrating Incremental Distributed Simulation Cloning

Figure 9.8 illustrates a simple supply-chain simulation comprising three federates, namely *simAgent (SA)*, *simFactory (SF)*, and *simTransportation (ST)*. A cloning trigger is predefined for federate *simFactory*, which contains a cloning condition "*OrderSize > MAX?*" and four candidate policies. The simulation models the supply chain operation of one-year duration. The *simFactory* reports the cost incurred in each order and in the whole year at the end of the simulation. Chapter 10 gives a detailed description of this example.

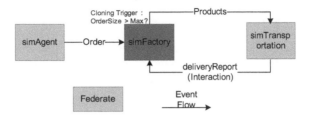

FIGURE 9.8
A distributed simulation example.

Figure 9.9 depicts the simulation execution using incremental simulation cloning to examine three of the candidate policies. Each scenario is marked as *Scen[i]* (i = 0, 1, 2), in which *Scen[0]* denotes the initial scenario. The incremental simulation cloning occurs along the time axis as follows:

At time 0, the simulation is initialized with a single scenario *Scen[0]*. When simulation progresses to time T_1, *SF[0]* performs active cloning due to an order with extra large volume, which results in the creation of clones *SF[1]* and *SF[2]*, and new scenarios *Scen[1]* and *Scen[2]*. The remaining federates do not need to be cloned immediately, and they only need to expand their associated region to enable them to continue interacting with *SF[1]* and *SF[2]*. Thus *SA[0]* and *ST[0]* become shared clones in both scenarios. The event flow from *SF[i]* (i = 0,1,2) to *ST[0]* is named *ev_F[i]* (i = 0,1,2). *ST[0]* keeps intact as long as *ev_F[i]*(i = 0,1,2) remain identical.

At simulation time T_2, *ev_F[0]* deviates from *ev_F[1]* and *ev_F[2]*; this triggers a passive cloning of *ST[0]* and results in the creation of clones *ST[1]* and *ST[2]*. This passive cloning does not trigger any change in existing scenarios. *SA[0]* persists as a shared clone after that.

Using the incremental cloning algorithm, clones are created as required according to the changing external conditions. For the whole simulation session, we always have:

Total no. of federates $\leqslant \sum$*No. of federates executing in each scenario*. For example, as shown in Figure 9.9, from simulation time T_0 to T_1 there exist only five federates simulating three scenarios whereas there must be nine federates examining the same scenarios in the context of traditional distributed simulations or using the entire cloning approach. Both incremental and entire cloning avoid repeating the computation of the original scenario before cloning. However, the incremental cloning approach also enables the sharing of computation among independent coexisting scenarios after cloning.

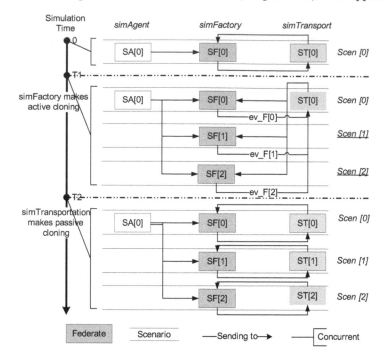

FIGURE 9.9
Executing simulation with incremental cloning.

9.5.2 Managing Shared Clones

A shared clone is capable of operating in multiple scenarios as long as it keeps receiving identical events from all scenarios in which it participates. The shared clone persists in this mode until the condition of triggering passive cloning is met. Thus during this time, the computation of this clone can be shared by different scenarios. The incremental cloning mechanism aims to make full use of the interdependencies among related scenarios, which is supported by a sensitive event-checking algorithm.

Sensitive events are those received from sibling clones in different scenarios, and sensitive events are comparable if they are updates of the same original object instance or interactions belonging to the same class (see Chapter 6 for details). Sensitive events are checked by the Callback Processor which is one part of the RTI++ middleware built upon the Decoupled federate architecture (see Chapter 3). Figure 9.10 illustrates the primary elements inside the Callback Processor, designed for checking events. The Sensitive Event Checker checks events and invokes the Control Module to trigger passive cloning when necessary. Mapping Tables maintain the relationships among scenarios and federates and the object instances (original/image) registered by related sibling clones. These tables are established and updated during the state replicating

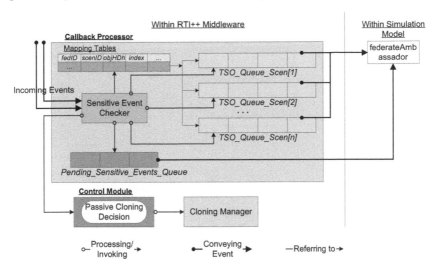

FIGURE 9.10
Internal design of the callback processor.

procedure on cloning. The checker can identify the source clone and scenario of each event via the tables; thus it can verify which events are comparable. A queue *Pending_Sensitive_Events_Queue (PSEQ)* stores the **target-sensitive** events with which other incoming sensitive events must be compared. The queue can be either empty or contain events with the same timestamp at any point in the simulation; this timestamp is referred to as the characteristic timestamp of *PSEQ*. A set of TSO event queues, *TSO_Queue_Scen[i]*, abbreviated *TQS[i]* ($i = 1, 2, \ldots,$ n), is established to buffer the events from each scenario in the corresponding queue. Events in those queues can be presented to the simulation model as appropriate.

The event-checking algorithm determines which events and how these events should be conveyed to the simulation model. The event checking decides whether or not a passive cloning is required and at which point the cloning should be triggered. A shared clone is said to be in **pending-passive-cloning** mode during the interval from deciding that a passive cloning is required to carrying out the cloning. Event checking is performed in the Callback Processor when control of a federate process is still with the RTI. Thus cloning will only be carried out when the RTI returns control to avoid potential problems incurred by replicating a federate while the RTI invokes callbacks. Figure 9.11 illustrates the algorithm of checking sensitive events by the Callback Processor, which are as follows:

- **Testing sensitive event**. A received update/interaction is tested according to definition 1 and 2 in section 6.3.2. In the case where the event

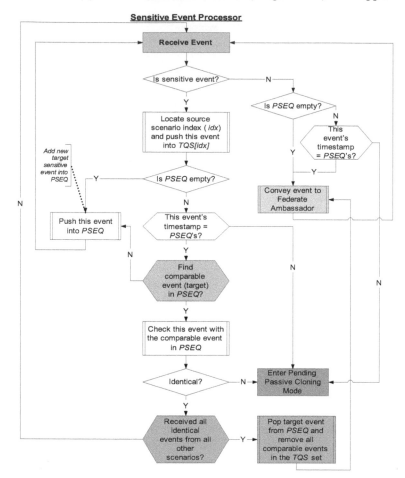

FIGURE 9.11
Checking sensitive events.

is a sensitive one, the checking continues. The processing of non-sensitive events will be covered later.

- **Identifying event source.** Mapping Tables are referenced to locate the source of this event. Hence the event checker can enqueue this event into the corresponding *TQS* queue.

- **Checking pending sensitive event queue.** If *PSEQ* is empty, the event checker pushes this event into *PSEQ* and sets its characteristic timestamp equal to the event's, after which current processing ends. When *PSEQ* is not empty, the event checker compares its characteristic timestamp with the event's. In the case where they are not equal (event's timestamp >

PSEQ's), the shared clone will enter pending-passive-cloning mode; otherwise the processing continues.

- **Locating target comparable event**. The event checker searches *PSEQ* to locate the comparable event to the event being processed. If *PSEQ* does not contain any comparable event, the event will be pushed into *PSEQ* and the processing ends. Otherwise, the event checker checks if the received event and the target event in *PSEQ* are identical. If they are not identical, the shared clone will also enter pending-passive-cloning mode; otherwise the processing continues.

- **Checking the progress status of processing**. The event checker examines whether the shared clone has received identical comparable events from all other scenarios. If so, the event checker presents the target event from *PSEQ* to the Federate Ambassador and deletes the comparable events inside the *TQS* queue set. If not, then the event checker waits for the next event.

The last step ensures the Federate Ambassador receives/reflects only one single event for one full set of identical events obtained from all related scenarios. This design hides the complexity of checking events from multiple scenarios. As a result, shared clones operate in multiple scenarios as if they only interact with one single scenario independently.

In the case where a non-sensitive event is received, the PSEQ's characteristic timestamp also must be compared when *PSEQ* is not empty. If the event's timestamp is greater than the characteristic timestamp, the shared clone requires passive cloning. This is because the shared clone will no longer receive identical events from any of the related scenarios to the target events in *PSEQ*. In the case that their timestamps are identical or *PSEQ* is empty, the event checker delivers this event to the Federate Ambassador directly. If *PSEQ* contains target-sensitive events, the decision of triggering a passive cloning depends on both the incoming events and the next granted time. Once the Callback Processor gets a granted time greater than (or equal to) PSEQ's timestamp[2], the shared clone enters pending-passive-cloning mode.

---

[2]If the granted time is equal to PSEQ's timestamp, setting pending-passive-cloning or not depends on whether the shared clone requests the last time advance by calling timeAdvanceRequest (TAR) / nextEventRequest (NER) or timeAdvanceRequestAvailable (TARA) / nextEventRequestAvailable (NERA).

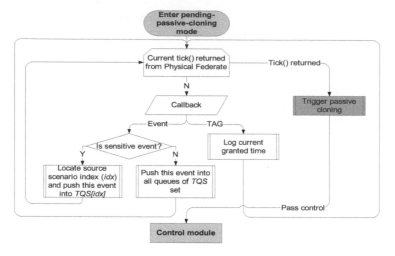

FIGURE 9.12
Processing events in pending passive cloning.

When a shared clone is in pending-passive-cloning mode, the Callback Processor buffers the incoming events as illustrated in Figure 9.12. Sensitive events should be enqueued to the corresponding *TQS* queue, whereas non-sensitive events should be inserted into all *TQS* queues unselectively. Thus, when new clones are created later on, each clone can straightforwardly inherit the events belonging to the scenario in which it operates. The Callback Processor logs the *timeAdvanceGranted* callback and the granted time. All callbacks are retained and not delivered to the simulation model until the pending cloning has been completed. Such a design aims to keep the semantics of the HLA specification and minimize the complexity of dealing with potential callbacks during pending-passive-cloning.

9.6 Summary

In this chapter, we introduced the infrastructure for HLA-based distributed simulation cloning. Our design enables distributed simulation cloning using a Decoupled federate architecture. The Cloning Manager module was developed to ensure correct cloning of distributed simulations when preset conditions are met at Decision points. During cloning, coordination and synchronization are required to maintain the state consistency. We presented the mechanism and algorithms for managing distributed simulation cloning, including active cloning and passive cloning.

The incremental cloning mechanism initiates cloning only when strictly necessary. Clone sharing among multiple scenarios is supported by a sensitive event-checking algorithm. The algorithm facilitates accurate sharing of clones and delays the passive cloning as long as possible. An entity mapping approach is designed to identify events, which maps the new HLA entities created for the clones in state replication to the existing ones known to the simulation models. The proposed incremental cloning mechanism supports correct HLA semantics and user transparency.

10

Experiments and Results of Simulation Cloning Algorithms

CONTENTS

10.1   Application Example

This chapter presents a simulation example (see Figure 10.1) to verify the correctness and investigate the performance of the proposed distributed simulation cloning technology. The example studies a simple supply chain that comprises an agent company, a factory, and a transportation company. The agent keeps issuing orders to the factory, and the latter processes these orders and plans production accordingly. The transportation company is responsible for delivering products of the factory and reporting the delivery status. The factory has a limited manufacturing ability, which means it needs to adjust its daily operation policy in the case of fulfilling an extra large order. One of the following candidate policies can be chosen:

- Keeping normal operation policy (NORMAL policy); thus the factory may get penalized because the order is not fulfilled in time.

- Encouraging overtime work to ensure on time delivery (OVERTIME policy); the factory has to pay for the extra manpower.

- Sharing a partial order with other competing factories (SUBCONTRACT policy); this also incurs profit loss and negative impact on future business development.

- Expanding manufacturing ability by purchasing assembly lines and recruiting more workers (EXPAND policy); thus extra cost will be incurred.

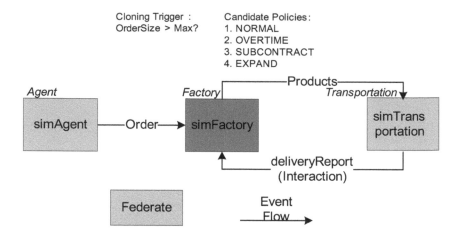

FIGURE 10.1
A simple distributed supply-chain simulation.

The analyst focuses on optimizing the extra cost/profit loss of the factory operation in processing large orders. Four different scenarios must be constructed to examine the candidate policies.

10.2   Configuration of Experiments

The three nodes in the supply-chain can be modeled as three federates as shown in Figure 10.1, namely *simAgent, simFactory* and *simTransportation*. These federates form a simple distributed supply-chain simulation. Two object classes, Order, Products and one interaction class, deliveryReport are defined in the FOM [66] to represent the types of events exchanged amongst the federates. Table 10.2 gives the classes published and/or subscribed by the federates. The *simFactory* reports the extra cost incurred for each order and for the whole year (from simulation time 0 to 361) at the end of the simulation. The experiments use four computers in total (three workstations and one server), which are interlinked via a 100Mbps-based backbone. Each federate (together with its clones if any) occupies one individual computer, with the RTIEXEC and FEDEX processes running on another computer.

The experiment architecture and platform specification are listed in Table 10.2. Using the same codes for the simulation models, the federates are built into three different versions by linking to (1) the DMSO RTI library directly (TRADITIONAL), (2) an RTI++ middleware library supporting entire cloning (CLONING_ENTIRE), and (3) an RTI++ middleware library supporting incremental cloning (CLONING_INCREMENTAL). We can also configure the federates of the 2nd (or the 3rd) version to execute each of the policies without using cloning (CLONING_DISABLED) to examine the correctness of using the decoupled architecture to build federates. The simulation execution using incremental cloning is discussed in Section 9.5.1 and illustrated in Figure 7.9.

TABLE 10.1

Declaration Information of the Federates

Federate	Object Classes and Attributes		Interaction Classes and Parameters
	Order	Products	deliveryReport
	Index,Size	*Amount,Index,Date*	*Index,Status*
simAgent	Publish	NIL	NIL
simFactory	Subscribe	Publish	Subscribe
simTransportation	NIL	Subscribe	Publish

10.3 Correctness of Distributed Simulation Cloning

The ten sets of experiments in total are indicated in Table 10.3. To verify the correctness of the cloning mechanism, we specify federate *simAgent* to generate the same set of orders in different runs. We first execute the TRADITIONAL federates, in which the collected results are used as a reference in subsequent experiments. Second, we repeat these experiments using CLONING_DISABLED federates. The last two experiments adopt CLONING_ENTIRE and CLONING_INCREMENTAL federates to explore multiple policies in each of the sessions. The outputs obtained using TRADITIONAL federates are summarized as follows:

- *simAgent* issues 240 orders, in which the first extra large order and the last order carries timestamp 83 and 362.5, respectively (note the last order is not received as it is after the simulation end time).

- *simFactory* receives 239 orders and makes products according to the

TABLE 10.2

Configuration of Experiment Test Bed

Specification	Computers				
	Workstation 1	Workstation 2	Server 1	Workstation 3	Workstation 4~13
Operating System	Sun Solaris OS 5.8	Sun Solaris OS 5.8	Sun Solaris OS 5.8	Win2000 Profes- sional	Sun Solaris OS 5.9
CPU	Sparcv9 CPU, at 900 MHz	Sparcv9 CPU, at 900 MHz	Sparcv9 CPU*6 at 248 MHz	Intel 1700 MHz Pentium IV	Sparc II CPU, at 400MHz
RAM	1024M	1024M	2048M	256M	512M
Compiler	GCC 2.95.33	GCC 2.95.3	GCC 2.95.3	MS VC++ 6.0	GCC 2.95.3
Underlying RTI	DMSO NG 1.3 V6	DMSO NG 1.3 V6	DMSO NG 1.3 V6	DMSO NG 1.3 V6	DMSO NG 1.3 V6
Processes running on	simAgent	SimTransp -ortation	simFact -ory	RTIEXEC & FEDEX	SimAgent or simTrans -portation

policy adopted. Different costs are reported in experiments *T_Norm*, *T_Over*, *T_Sub*, and *T_Exp*, respectively.

- *simTransportation* receives all product updates issued earlier than the end time and sends *deliveryReport* interactions with respect to these updates.

In order to evaluate the correctness of the cloning mechanisms, we examine the orders issued and received, products produced and delivered as well as the costs. We have the following observations from the experiments using federates linked with the other libraries:

- Outputs in the experiments with CLONING_DISABLED federates (*D_Norm*, *D_Over*, *D_Sub*, and *D_Exp*) match exactly those using TRA-DITIONAL federates.

- In experiments Ec_Multiple, *simFactory* makes one, two or three new clones at simulation time 83 to execute two, three, or four policies concurrently. Immediately after this, passive cloning is forced in both *simAgent* and *simTransport*. For example, in the experiment to examine three policies, after cloning there are nine federates in total to explore three scenarios

TABLE 10.3

Experiments for Verifying the Correctness of Cloning Technology

Type of Federates	Candidate Policies			
	NORMAL	OVERTIME	SUBCONRACT	EXPAND
TRADITIONAL	T_Norm	T_Over	T_Sub	T_Exp
CLONING_ DISABLED	D_Norm	D_Over	D_Sub	D_Exp
CLONING_ ENTIRE	Ec_Multiple			
CLONING_ INCREMENTAL	Ic_Multiple			

concurrently. Each clone of *simFactory* reports exactly the same results to those in *T_Norm*, *T_Over*, *T_Sub*, and *T_Exp*, respectively.

- In experiments Ic_Multiple, *simFactory* makes clones in the same way as in the *Ec_Multiple* experiments. The difference is that federate *simAgent* keeps intact all the time and *simTransportation* remains shared until simulation time 101 when it performs passive cloning, required by the event checking algorithm (see section 9.5). In the experiment to examine three policies, there are five federates from time 83 to 101 while there are seven federates after time 101 to explore three scenarios (see Figure 7.9). Each clone of *simFactory* reports exactly the same results to those in *T_Norm*, *T_Over*, and *T_Sub*, respectively.

Outputs of the above experiments indicate that the technology provides a correct cloning mechanism for HLA-based distributed simulations. This also proves that the Decoupled federate architecture ensures complete fidelity in the way it bridges the simulation model and the Local RTI Component. In other words, the simulation cloning technology does not introduce any variation in the simulation results.

10.4 Efficiency of Distributed Simulation Cloning

To investigate the performance of the cloning technology, we carry out a series of experiments to collect the overall execution time using federates linked to different libraries to execute all policies (Table 10.4). For traditional federates, we execute the policies one by one, and for each policy we carry out a number of runs with the extra large order generated randomly at any time (ranging from the start time to the end time) in each run. The execution times of all runs are averaged, and the result is referred to as the average time of executing one single scenario. As for experiments with cloning-enabled federates, we let federate *simAgent* generate different orders. We select three runs in which the extra large order occurs at time 80, 203, and 320, thus, federate *simFactory* may trigger active cloning at different stages in each run. These points represent cloning at the start, middle, and end stages, respectively. Furthermore, we also specify *simFactory* to make a different number of clones to examine alternative policies in different experiments (2, 3, or 4, policies respectively).

TABLE 10.4
Experiments for Studying the Efficiency of Cloning Technology

Type of Federates	Experiments								
Cloning Stage	Start			Middle			End		
Number of Policies	2	3	4	2	3	4	2	3	4
CLONING _ENTIRE	Ec_ s_2	Ec_ s_3	Ec_ s_4	Ec_ m_2	Ec_ m_3	Ec_ m_4	Ec_ e_2	Ec_ e_3	Ec_ e_4
CLONING _INCREM -ENTAL	Ic_ s_2	Ic_ s_3	Ic_ s_4	Ic_ m_2	Ic_ m_3	Ic_ m_4	Ic_ e_2	Ic_ e_3	Ic_ e_4

The average CPU utilization of a single traditional federate (in workstation 1 or 2) is reported as above 80%. In the case of enabling simulation cloning, the average CPU utilization of each clone is reported as ~44%, ~30%, or ~21%, respectively, when there are two, three, or four clones running on a single workstation. Physical federates have an average CPU utilization as low as ~1%. Experimental results are recorded in seconds in Figure 10.2. The average time for executing one single scenario per run is 561 seconds using traditional federates. The percentage of saved execution time using cloning technology is shown in Figure 10.3. This uses the execution times reported by traditional federates running scenarios sequentially as a reference.

Figure 10.2 shows that the cloning-enabled federates can significantly reduce execution time compared with traditional federates. The experimental results indicate that the more Computation there is in common among different scenarios, the more execution time can be reduced using simulation cloning. It also shows that the larger the number of scenarios to be examined using cloning, the more execution time can be reduced. Results in Figure 10.3 indicate that the incremental cloning approach has an obvious advantage over the entire cloning approach in terms of execution efficiency under all given configurations. This is because the incremental cloning approach can further save Computation by supporting federate sharing among scenarios.

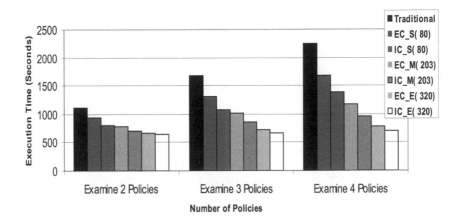

FIGURE 10.2
Execution time for examining multiple scenarios.

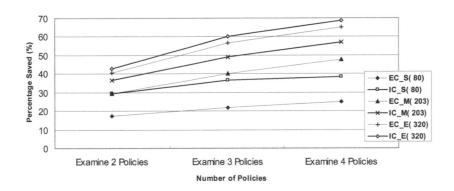

FIGURE 10.3
Percentage of saved execution time using entire and incremental cloning.

10.5   Scalability of Distributed Simulation Cloning

Another series of experiments are performed to test the scalability of the cloning technology. This consists of six sets of experiments in total, with the initial number of federates varying from three to thirteen. Each set of experiments uses a federation that always contains one *simFactory* and one or multiple instances of *simAgent* and *simTransportation*. In the nth set of experiments, the federation initially has $2n+1$ federates as shown in Figure 10.4. Similar to the previous experiments, the execution time for examining the first three policies (see Section 10.1) is measured and compared using different types of federates (TRADITIONAL, CLONING_ENTIRE, and CLONING_INCREMENTAL). The experiment architecture and platform specification for the scalability study are listed in Table 10.2. Federates *simAgent.1~n* are configured to issue extra large orders at the middle stage of the simulation. The first extra large order received by the *simFactory* acts as a trigger of an active cloning, but after cloning, each particular scenario applies to all subsequent orders, including extra large ones.

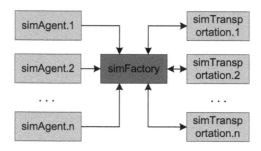

FIGURE 10.4
Initial federation for the nth set of experiments for the scalability test.

The average CPU utilization of each federate is the same as in previous experiments. The experimental results are recorded in Figure 10.5. The sum of the execution times using traditional federates increases smoothly when the federation contains an increasing number of federates (starting from 1683 seconds for three federates to 2,169 seconds for thirteen federates), and the same trend can be observed when using cloning enabled federates. Figure 10.5 indicates that cloning technology can reduce the execution time significantly. Figure 10.6 shows that using incremental cloning the percentage of saved execution time is about 47% to 49%. The entire cloning approach gains about 30% to 39% time saving. The experimental results show that the proposed cloning technologies scale well with increasing distributed simulation size, and the incremental cloning approach consistently performs better than the entire cloning approach.

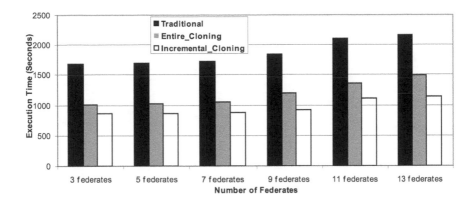

FIGURE 10.5
Execution time for examining three policies with increasing number of federates.

10.6 Optimizing the Cloning Procedure

The simulation Cloning procedure is defined as the interval from the point where the parent federate starts cloning by calling the *createClone* method (or *splitClone* in the case of passive cloning) to the point where all new clones complete initialization, the remaining federates are ready for resuming normal execution, and control is returned to the simulation models (see Figure 7.3 and Figure 7.6). The cloning algorithm requires the following RTI-related operations to be performed in sequence during the cloning procedure (as described in Section 9.2):

- Save the federation.

- For each new clone, join the existing federation.

- Synchronize the federation.

Such a simulation cloning procedure takes about 20 seconds in the previous experiments. This overhead is mainly due to the above operations that incur costly federation-wide coordination, especially the *joinFederationExecution* call. In DMSO RTI-NG, this operation requires opening TCP sockets to all other federates in the federation, which is expensive [95]. Although the time taken by these operations will increase with the number of federates, the exact relationship will depend on the internal design of these RTI services, which is transparent to the user. It is therefore worthwhile to optimize the cloning procedure. A possible solution is to avoid the *joinFederationExecution*

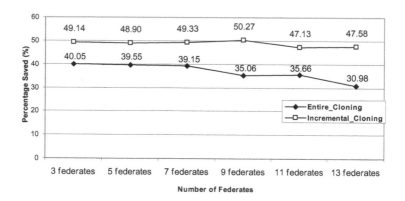

FIGURE 10.6
Percentage of saved execution time with increasing number of federates.

call during the cloning procedure itself. We attempt to solve this problem using a physical federate (**phyfed**) pool approach as shown in Figure 10.7. This approach creates phyfed instances concurrently with the normal simulation execution, which can potentially serve the new clones in the future.

Before cloning occurs, several existing phyfed instances are created, which join the federation and form a phyfed instance pool. In the context of the pool approach, a phyfed instance may operate in two modes: (1) **idle mode**, calling tick regularly to maintain connection session with the RTI while checking for invocation from a virtual federate; and (2) **working mode** Working mode, servicing a virtual federate as normal. An idle phyfed instance is neither time regulating nor constrained, and has minimum interaction with other federates.

When cloning occurs, a phyfed instance can be fetched from the pool by the virtual federate to provide the required RTI services immediately. Thus, this approach avoids new clones taking part in the *joinFederationExecution* prior to state replication. However, maintaining spare phyfed instances will incur extra overhead in that (1) each instance consumes system resources and (2) the RTI has to monitor the session with more federates.

In order to measure the performance of this approach, a series of experiments is carried out. The configuration and structure of the experiments are similar to the supply-chain example and are given in Figure 10.8 (see Table 10.2 for the platform specification). The simulation contains three federates (*Fed[1] to [3]*), which are built into two different versions by linking to (1) the previous RTI++ middleware supporting incremental cloning and (2) an RTI++ middleware supporting the phyfed pool approach. *Fed[2]* is configured to perform active cloning (spawning 1*sim*3 clones in different runs) when it is granted federate time 100, and this leads to passive cloning of *Fed[3]* after-

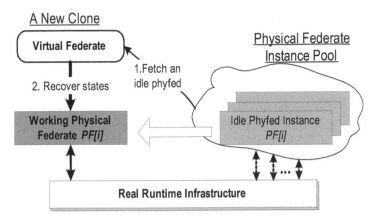

FIGURE 10.7
Physical federate pool approach

ward. The average CPU utilization of each federate is the same as in previous experiments.

In these experiments, we compare the initialization time of federates, the time for the cloning procedure, and the overall execution time using the two different versions of middleware. The initialization time denotes the interval from when a federate is started to the point immediately after it completes the following operations: join the federation, enable time regulating/constrained, publish/subscribe object/interaction classes, and register object instances. The phyfed pool is created at the time when the original federate invokes *createFederationExecution*, and a different number of idle phyfed instances are created according to the number of potential policies. Table 10.6 lists the results of the experiments.

TABLE 10.5
Comparison between Normal and Pool Approach

Number of Clones	Idle Phyfed In- stances	Initialization Time (seconds)		Cloning Time (seconds)		Execution Time (seconds)	
	For Each Federate	Normal	Pool Approach	Normal	Pool Approach	Normal	Pool Approach
1	1			∼20	∼3	∼93	∼83
2	2	19∼27	21∼27	∼23	∼3	∼117	∼107
3	3			∼23	3∼4	∼140	∼131

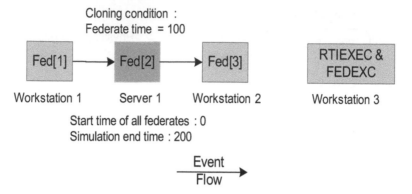

FIGURE 10.8
Experiments for studying physical federate pool approach.

The initialization time in the pool approach is close to the normal one, which indicates that the overhead of phyfed instances joining the federation is hidden by the other initialization operations. The time used for cloning is dramatically reduced from ~20 seconds to ~3 seconds, and most of the time is spent in the federation save operation. Using the pool approach can reduce the overall execution time compared with the normal cloning approach. However, the improvement in cloning time is reduced by the overhead for maintaining idle phyfed instances. To achieve the best performance using this approach, phyfed instances should be added to the pool close to the point of cloning. In this way, the reduction in cloning time can best outweigh the redundancy of idle phyfeds. Another concern is to optimize the time consumed in the federation save operation, which is only used to ensure that in-transit messages will arrive at destinations prior to cloning. An alternative candidate approach for dealing with in-transit messages is introduced in Chapter 3.6.

10.7   Summary of Experiments and Results

In this chapter, an example of distributed supply chain simulation was used for examining the correctness, performance, and scalability of the distributed simulation cloning technology. Experiments were carried out to compare two cloning mechanisms, namely entire cloning and incremental cloning, in terms of execution time using normal federates as a reference.

The simulation outputs demonstrated that the technology provides a correct cloning mechanism for existing distributed simulations. The performance results showed that the technology can reduce the time of executing multiple scenarios significantly. The results also showed that the proposed cloning technology has a promising scalability. Furthermore, incremental cloning can provide further savings by sharing the Computation of alternative scenarios compared with an approach using entire cloning.

10.8   Achievements in Simulation Cloning

We have investigated important research issues in HLA-based distributed simulation cloning and have presented alternative management mechanisms. Simulation cloning offers the flexibility for a distributed simulation to examine multiple decision policies concurrently at a decision point. Thus, the execution time can be reduced and the analyst can quickly obtain multiple sets of results that represent the impacts of alternative decisions. The technology provides simulation users with a decision-support platform much more powerful and flexible than the traditional linear approach. All the objectives set at the beginning of the book have been fulfilled. The major achievements are summarized as follows:

- Establishment of the foundation theory of distributed simulation cloning.

- Investigation and design of efficient and reliable solutions to support distributed simulation cloning.

- Development of a generic approach for federate cloning using our decoupled federate architecture.

- Development of mechanisms for management of concurrent scenarios in cloning-enabled simulations.

- Design and development of RTI++ middleware to provide reusability of existing simulation models and user transparency.

- Investigation and development of alternative distributed simulation cloning mechanisms, including entire and incremental cloning.

This book establishes the foundation theory of distributed simulation cloning, defining the terms and identifying the issues involved. A federate can perform cloning actively or passively according to its own or its partner's requirements. When cloning of a federate is required, an entire cloning mechanism replicates the whole scenario, whereas an incremental mechanism only clones those federates whose states will alter at the decision point. Execution of identical federates (clones) may be shared by multiple scenarios. Thus, there exist dynamic relationships among scenarios and federates.

We consider Multiple-Federation (MF) and Single-Federation (SF) solutions to distributed simulation cloning. This book compares and analyzes their advantages as well as drawbacks. The Multiple-Federation solution has the advantage of robustness and provides simple synchronization of federates in each individual scenario. Using the Single-Federation solution, management of cloning and Computation sharing among scenarios is much easier to achieve.

These candidate solutions have been analyzed and compared in terms of efficiency, complexity in management, and reliability. In the case where there is high data exchange, the MF solution can perform much better than the simple SF solution as the concurrent scenarios are not efficiently partitioned when the latter solution is applied. The two solutions exhibit close performance by applying Data Distribution Management services to the Single-Federation solution to give the DSF solution. Furthermore, the two solutions have no significant difference in terms of federate synchronization. As DSF is also superior in convenience for design and scenario management, it has been adopted for developing the infrastructure of distributed simulation cloning.

In order to tackle the problems of replicating a federate instance and providing fault tolerance to HLA-based distributed simulations, a Decoupled federate architecture has been designed. This approach decouples the simulation model from the Local RTI Component, with these two basic modules forming a single federate executive. A virtual federate process executes the simulation model while a physical federate process provides RTI services to the corresponding model at the backend. The decoupled approach successfully supports user transparency using the standard HLA specification to provide an interface to the simulation model. The approach maintains the semantics of executing RTI service calls and conveying RTI callbacks to the simulation model. The decoupled architecture facilitates state saving and replication at the RTI level. It has been used to develop a mechanism for replicating the simulation federate, which is reliable, correct, and relatively model independent.

Decoupled federates have been compared with normal federates in terms of latency and time advancement performance. The decoupled architecture incurs a negligible latency overhead for small or medium payloads and only a slight extra latency in the case of a bulky payload. Furthermore, the Decoupled federate architecture has a promising performance of time advancement to normal federates. This makes the architecture an ideal approach to support cloning. Concurrent scenarios coexist in a single federation session as a result of simulation cloning, each representing a particular execution path. Interactions must be confined within each scenario to guarantee the correctness of simulation results. On the other hand, a mechanism to share Computation is required by the shared clones that cross multiple scenarios. DDM services have been used at the backend to route events and partition scenarios. In this book, we have studied issues involved in managing concurrent scenarios and suggested candidate approaches. To obtain the best DDM performance, the internal mechanism of DDM services implemented in DMSO RTI-NG has been analyzed.

With the spawning of scenarios, the number of clones in a cloning-enabled simulation session keeps increasing. We have developed the idea of a scenario tree to identify and denote scenarios correctly and efficiently. In this book, we discussed two possible solutions that we have designed to code scenarios and to manipulate characteristic region extents associated with a scenario, namely a recursive region division solution and a point region solution. The former solution has advantages in minimizing the complication for sharing federates (clones) and mapping the scenario identity to the region extent. However, the point region solution has been adopted in our cloning technology because it can meet the region allocation requirements in any situation while optimizing the use of networking channels for DDM services. The scenario management mechanism ensures both proper scenario partitioning and accurate Computation sharing across scenarios, which are potentially contradictory requirements. The mechanism also accurately represents and manipulates the dynamically complex relationship among increasing numbers of scenarios.

A middleware approach is used to hide the complexity incurred by simulation cloning. A cloning-enabled infrastructure, namely RTI++, contains the modules for controlling, and managing cloning, managing scenarios as well as handling cloning regions and the underlying DDM services. The RTI++ presents a standard RTI interface to the simulation model. The infrastructure supports the correct HLA semantics and user transparency, which maximizes the reusability of legacy simulation applications and also minimizes the users' efforts to utilize cloning technology in developing their models.

One important goal of cloning technology is to optimize execution by avoiding repeated Computation among independent scenarios. Alternative cloning mechanisms have been developed: an entire cloning mechanism and an incremental cloning mechanism. We have investigated and designed algorithms for managing active and passive cloning, which ensure correct replication of federates when required. These algorithms maintain the state consistency of the simulation session by introducing proper coordination and synchronization on cloning. State saving and replication are designed using the Decoupled federate architecture.

Although the incremental mechanism requires a more complex control algorithm, it makes the most use of Computation sharing. In this book, we have studied the incremental cloning mechanism and established a theory of incremental distributed simulation cloning. The incremental mechanism initiates cloning of a federate only when strictly necessary, while the entire cloning mechanism makes replicas of every federate in a scenario. A sensitive event checking algorithm facilitates clone sharing across scenarios, and avoids/delays the passive cloning as long as possible while guaranteeing the correctness of Computation sharing. The checking algorithm employs an entity mapping approach to identify events, which maps the new HLA entities created for the clones in state replication to the existing ones known to the simulation models, and vice versa. The incremental cloning mechanism not only avoids repeated Computation prior to cloning, but also further saves identical federate execution among scenarios without prior knowledge of the behavior of the federate.

A series of experiments has been performed to investigate the correctness and performance of the two alternative mechanisms using an example of a distributed supply chain simulation. The experimental results are compared for traditional, cloning-disabled, and cloning-enabled federates in terms of the consistency of outputs and computing efficiency. The experimental results show that the technology provides a correct cloning mechanism for HLA-based distributed simulations. The experimental results also show that the cloning technology can reduce the time of executing multiple scenarios of distributed simulations. The scalability of the cloning mechanisms has also been examined, and experimental results show that the proposed cloning technology has promising scalability for both mechanisms.

However, there is still not a general solution to the problem of a possible combinatorial explosion of scenarios in distributed simulation cloning in some extreme situations. Although it is unlikely to occur, simulation analysts should be aware of the problem and avoid defining too many cloning triggers or specifying a large number of new scenarios on active cloning.

The Decoupled federate architecture can also be exploited in providing fault tolerance and load balancing as well introducing Web and Grid technologies to HLA-based simulations. In this book, we have discussed and proposed candidate solutions to address these issues. An alternative scheme for dealing with in-transit messages was also described. The HLA-based distributed simulations may benefit from the Decoupled federate architecture in the following ways:

- **Fault tolerance.** We can easily isolate the faults of part of a federation and provide fault recovery without rolling back the simulation execution.

- **Web-or Grid-Enabled Architecture.** The decoupled architecture can be exploited to combine the advantage of Web/Grid services and the HLA. HLA-based simulations may profit from the flexible resource management and enhanced interoperability provided by a Web/Grid-enabled architecture.

- **Supporting load balancing.** Using the decoupled architecture can free developers from handling RTI, states and in-transit events, and it avoids the overhead incurred in saving/restoring RTI states when adopting a federate migration approach to balance the federates load.

Part IV

Applications

11

Hybrid Modeling and Simulation of Huge Crowd over an HGA

CONTENTS

11.1 Introduction

Modeling and Simulation are at the very core in many areas of science and engineering. With the rapid growth of both the complexity and the scale of problem domains, it has become a key requirement to create efficient and ever more complex simulations of large scale and/or high resolution for research, industry, and management. Modeling and simulation of crowd is a typical paradigm. As a collective and highly dynamic social group, a human crowd is a fascinating phenomenon, which has been constantly concerned by experts from various areas. Recently computer-based modeling and simulation technologies have emerged to support investigation of the dynamics of crowds.

Crowd modeling and simulation has now become a key design issue in many fields including military simulation, safety engineering, architectural design, and digital entertainment.

To represent the behavior of a crowd, a number of behavior models have been proposed with various types of modeling approaches [25], such as particle system models [51], flow-based models [58], and agent-based models [16]. Despite the many existing research efforts and applications in crowd modeling and simulation, it is still a young and emerging area. Work on modeling and simulation of large crowds (consisting of thousands of individuals or even more), especially at the individual level, is still rare. Large agent-based systems, such as simulation of large crowds at the individual level, have long been placed in the highly computation-intensive world [16, 25]. Study on crowd phenomenon still suffers from a lack of (1) an effective modeling approach to cope with the size and complexity of a scenario involving a huge crowd and (2) an appropriate platform to sustain such large Crowd simulation systems.

In the past few years, there have been a lot of successful attempts of incorporating Grid technologies to foster large simulations over the Internet, such as the Grid-aware Time Warp Kernel [69] and the HLA_Grid_RePast framework [16, 21]. However, these simulation systems are only suited for executing coarse-grained models due to the limited network bandwidth between different administrative domains. Another issue is that only a few nodes of an administrative domain are accessible to the external users due to the existing security rules of most administrative domains. The third issue concerns reusability: Large simulation developers often already have a set of simulation models/components situating over their intranet. Only with the advent of our hierarchical Grid infrastructure [25] can (1) existing individual bundles of simulation models be linked to form a dedicated large simulation crossing the boundaries of previously independent simulation groups and the boundaries of the administrative domains, (2) fine-grained models still benefit from the advantages of Grid technology, and (3) users have a flexible solution to the reusability issue.

This study employs the hierarchical Grid infrastructure as an effective approach to addressing the two pending issues in large crowd modeling and simulation. The resultant simulation system is hybrid in nature: (1) to capture the individuality of pedestrians, agent-based models have been developed with each agent representing a pedestrian; (2) to describe the global dynamics of the whole (or a part of interest) crowd, computational models have been used; and (3) these heterogeneous models are linked together to allow studying the interaction dynamics of a crowd at both the individual and global scales.

The large Crowd simulation system had been deployed over two distinctive administrative domains located in China and the United States respectively. The simulated crowd has a size which is prohibitively large for traditional simulation techniques using a cutting-edge office desktop. Our approach successfully alleviates the bottleneck in the design and analysis of particularly large and complex scenarios like huge crowd. This study is one of the first to provide a solution to simulation of a crowd with models of multiple scales and types. The remainder of this chapter is organized as follows: Section 11.2 briefs the background and related work of crowd modeling and simulation. Section 11.3 recaps the hierarchical Grid simulation infrastructure. Section 11.4 describes the case study of exploring the dynamics of a huge crowd in an evacuation procedure. Section 11.5 concludes this work with a summary and proposals on future work.

11.2 Crowd Modeling and Simulation

As a collective and highly dynamic social group, a human crowd is a fascinating phenomenon, which has constantly been a concern of experts from various areas. A crowd may exhibit highly complex dynamics; in general, pure mathematical approaches or analytic models are not adequate in characterizing the dynamics of a crowd.

Recently, modeling and simulation technologies have been gaining tremendous momentum in investigating crowd dynamics. Various simulation architectures have been developed [25, 87]. To represent the behavior of a crowd, a number of behavior models have been proposed with different types of modeling approaches, such as flow-based models and agent-based models. To study or mimic the dynamics of a crowd, modelers have considered a number of physical factors, social factors, and psychological factors when characterizing crowds in their models. Crowd models may also concern different aspects of a crowd. Some work aims at the external characteristics of a crowd, such as appearance, poses or movement patterns, coordinated positions of individuals; and some other works focus on how a crowds social behaviors evolve over time upon some events.

Many open research issues are still very much in flux due to the complexity of individual and crowd behaviors [120]. Modeling of a crowd first needs to determine the scale (level of detail) of the model. The existing models are largely at two extreme levels: either model each individual as an autonomous agent equipped with some human-like behavior modules such as locomotion, perception, and decision making, or treat the crowd as a whole or a collection of homogeneous particles with limited or no cognitive features. With the fast development of computing technology, there seems to be a trend in Crowd simulation to model each individual as some kind of intelligent agent with attempts to incorporate more and more social and psychological factors into the agent behavior model. However, we believe that the common shortcoming of existing models is the absence of modeling the social group process and its impact on human behavior.

The interoperability and composability issue is largely ignored by existing research. Although there already exist a number of Crowd simulation

Crowd simulation models and modeling approaches, these models can hardly work with each other: They may operate at different levels of abstraction. This makes the communication between different models difficult. In a typical Crowd simulation, the crowd model needs to pass various information about individuals actions and also needs to understand the events in the simulated world so that the individuals can determine how to respond to these events. However, research in interoperable Crowd simulation models is still rare.

This study attempts to address the two pending issues via the hierarchical Grid simulation architecture. We aim to explore an approach to constructing simulations of huge crowd constituted by models of various scales and types; thus the dynamics of a huge crowd may be investigated at different levels in a manner more comprehensive than existing approaches do.

11.3 Hierarchical Grid Architecture for Large Hybrid Simulation

In [25], we presented the hierarchical Grid computing architecture for large-scale simulation. The architecture serves as a simulation infrastructure, which can (1) cross distributed administrative domains, (2) link multiple distributed simulation models into a large-scale simulation over Grids, and (3) reduce the communication overhead among simulation bundles. This section recaps the basics of a Grid system as well as the design and functionalities of the architecture.

11.3.1 Grid System Architecture

As shown in Figure 11.1, a production Grid typically contains a number of compute centers, which are linked by high-speed networking. A compute center in general is organized at two levels. A head node, which hosts several access services and resource management functions for a compute center, accepts incoming computing jobs and schedules them to local computing and data resources, which are termed here Compute Element (CE), and Storage Element (SE), respectively. Inside a data center, the head node, CEs and SEs are interconnected by a high-performance local area network (LAN).

11.3.2 HLA-Based Simulation Model

The High Level Architecture (HLA) is a technology for simulator interoperability and the de facto standard [66] for simulations over distributed computing platforms. The HLA defines software architecture for modeling and simulation, and is designed to provide reuse and interoperability of simulation models/components, namely federates. A collection of federates interacting with each other for a common purpose forms a federation; in the context of

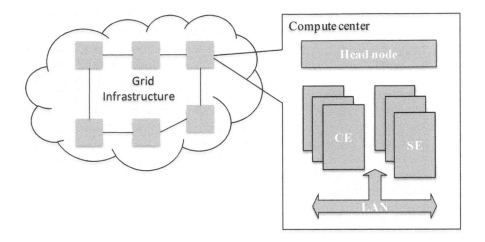

FIGURE 11.1
Grid computing system architecture.

this study, a simulation bundle is referred to as an independent federation running within a compute center. As the HLA defines the specification, it is the Runtime Infrastructure (RTI) that services federates for data interchange and synchronization in a coordinated fashion (see Figure 11.2). The RTI services are provided to each federate through its Local RTI Component (LRC) [18, 21]. The RTI can be viewed as a distributed operating system providing services to support interoperable simulations executing in distributed computing environments [51].

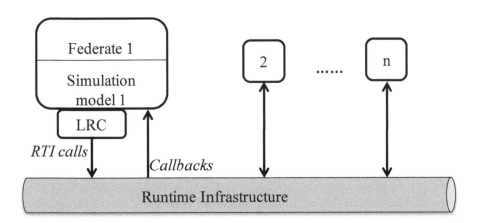

FIGURE 11.2
Model of HLA-based simulation.

11.3.3   Hierarchical Grid Simulation Architecture: Overview

The design of the hierarchical Grid Simulation Architecture adopts a concept model as shown in Figure 11.3. A number of federates are grouped into multiple simulation bundles. Inside a simulation bundle, multiple federates communicate with each other via the local RTI. Each simulation bundle hosts a gateway service, which coordinates the communication of federates from different simulation bundles. The concept model of hierarchical Grid simulation infrastructure was designed to conform to the Grid system defined in Section 11.3.1. As shown in Figure 11.4, a compute center hosts a simulation bundle; the gateway service is implemented in the head node, and federates of the simulation bundle are computed in CEs of the compute center. The gateway service communicates with gateways of other simulation bundles over the Grid infrastructure.

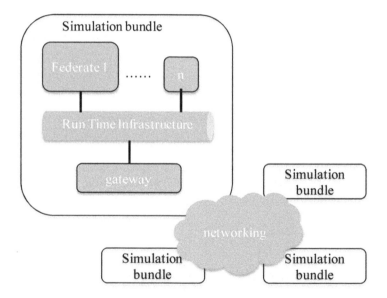

FIGURE 11.3
Concept model of the hierarchical Grid simulation architecture.

As illustrated by Figure 11.5, a gateway mainly consists of (1) a Grid federate module (GFM), (2) a Local federate module (LFM), (3) a Routing module, and (4) a Synchronization module. The GFM is implemented as a Grid service and communicates with other gateways. The GFM retrieves events and runtime simulation data generated by simulation models from other simulation bundles. The LFM operates an individual LRC to directly interact with federates in the simulation bundle. The LFM receives timestamped ordered (TSO) events and other runtime simulation data from its local federation and

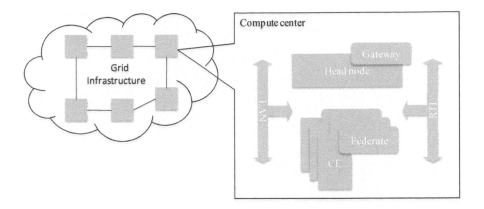

FIGURE 11.4
Large-scale distributed simulation in a hierarchical Grid architecture.

generates events from other federations (remote events) the local federation. The Routing module deals with delivering an event to the right destination gateway and relays remote events to the LFM. The synchronization module is responsible for delivering these events in correct order and coordinating time advances of all simulation models. More details about the synchronization algorithm are available in [20].

11.4 Hybrid Modeling and Simulation of Huge Crowd: A Case Study

We have developed a federated simulation of a huge crowds evacuation and its interaction with vehicles in an urban area. The objectives include (1) to examine the feasibility of composing models of various types to characterize a crowd, (2) to explore the suitability of the hierarchical Grid simulation architecture for Crowd simulation, and (3) to address the performance bottleneck of simulating a huge crowd.

11.4.1 Huge Crowd Scenario

We first identified a typical scenario of the huge crowd phenomenon: an evacuation procedure of a crowd after a National Flag Raising Ceremony at Tiananmen Square (see Figure 11.6(A)). The square is 440,000 square meters in area with a rectangular shape. The main body of the square is shown as the area

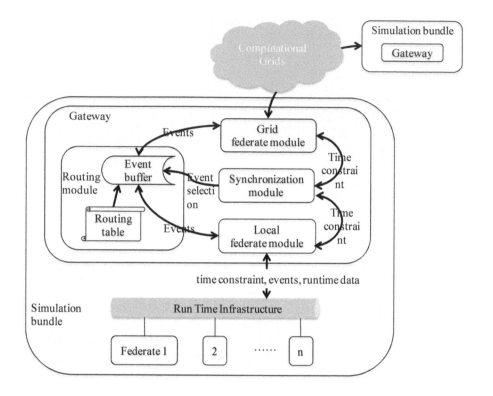

FIGURE 11.5
Gateway service architecture.

circled by the dashed line. Pedestrians may freely traverse in this area except
the zones restricted from access. Thousands of pedestrians can gather around
the zone of interest (flagged) and still remain in the assembly area with the
ongoing event. Figure 11.6(B) is a snapshot of the assembly area in real life.

On completion of the event, pedestrians start to disperse away from the
assembly area and leave the square via six passages, referred to as P1 through
P6. P1 and P2 denote the entrances of two pedestrian underpasses, and P3
through P6 represent four pedestrian crosswalks. As soon as pedestrians enter
P1 or P2, they will be considered out of the scenario. Two one-way roads located on the top and beneath the circled area respectively, and pedestrians can
leave the square to the other side of each road via the crosswalks (the reserve
direction not assumed) and leave the scenario afterward. When pedestrians
are moving on a crosswalk, the flow of vehicles along a road will be controlled
by traffic lights that signal every 60 seconds periodically. No violation of the
traffic rule is assumed for either pedestrians or vehicles.

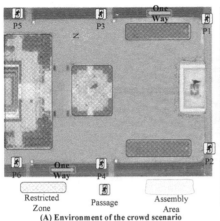

(A) Environment of the crowd scenario

(B) A case of the crowd in the assembly area

FIGURE 11.6
Illustrating the huge crowd scenario.

11.4.2 Simulation Models

To adapt to the size and complexity of the scenario, we developed three significantly different types of models with various scales and composed the huge Crowd simulation with them. Those include (1) an agent-based model for pedestrians in the square (except the assembly area) and the crosswalks, (2) a macroscopic crowd model for the pedestrians aggregated only in the assembly area, and (3) another agent-based model for the vehicles on the roads.

11.4.2.1 Pedestrian Agent Model

In this study, we adopted an agent model to characterize a pedestrian's behaviors. An agent represents a pedestrian, who can be an independent individual or a member of (1) a tourist group, (2) several friends or (3) a group of relatives. Under normal conditions, the model assumes that pedestrians would follow the rule of proximity; for example, pedestrians with their initial locations at the upper part of the square tend to select the upper exits as their final goals. Figure 11.7 depicts an agents state transition process spanning from its start state (initial location) to the end state (an exit reached) where it is removed from the simulation scenario.

An agent always tends to reach a passage to exit the scenario, and the passage will be selected as its final goal. We define a human-like rule for routing the agent to fulfill its final goal. The agent's route area is always divided into four connected areas, each associated with a sub-goal, that can be any location of the area. In a normal situation, that is, no danger event detected, the agent needs to check whether it is related to the neighbors. If

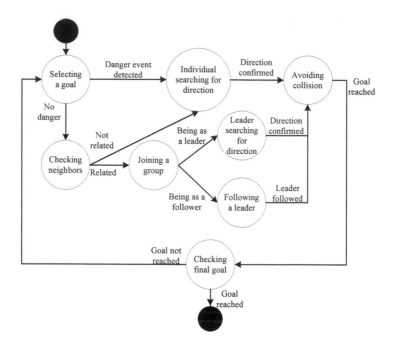

FIGURE 11.7
State transition of a pedestrian agent in terms of behaviors.

yes, the agent moves together with the related agent(s) as a group, and the leader agent in the group searches for direction while other agents follow the leader. In contrast, an independent agent searches for direction individually.

Once any agent identifies the direction, it switches to the behavior of avoiding neighbors. The agent checks from time to time whether the goal in the area is reached; if yes, it will move to the next area and do the check again; otherwise it continues searching for direction and avoiding behaviors until it reaches the final goal. If a danger event is detected, the agent directly searches for direction without considering the grouping option. Another possibility not covered by Figure 11.7 is that an agent can perish when encountering danger. The agent will reach the end state immediately and be removed from the scenario.

11.4.2.2 Computational Model of the Crowd Aggregated in the Assembly Area

When dealing with a crowd of high density and large size, a macroscopic model often applies. Such a macroscopic (computational) model treats a crowd as a whole to characterize some common features of the individuals in the crowd without distinguishing their individualities, for example, the flow model using the continuum theory [58]. A computational model can obtain results at the scale of a whole crowd at much lower computational expense in comparison with using agent-based models. In the study, we compose a computational model that defines the mechanism for the pedestrians in the assembly area (referred to as the crowd in this discussion), choosing the route to the destination and implementing the choice of routing: The crowd's speed is a dynamic three-dimensional function [114]. It defines the average speed of individuals located within a unit area of floor space at given a time point t and location (x, y). The expected number of these individuals is the pedestrian density, written as $\rho(x, y, t)$. The chosen speed and moving direction of these individuals, that is, the expected velocity, is written as $v(x, y, t)$. The crowd's speed is subject to the density. The speed monotonically decreases from a preferred speed down to zero, with the density varying from zero to a preset maximum value.

The model always chooses the route that requires the shortest time to reach the destination. We defined a cost function and a potential function for this purpose: The cost function, written as $c(x, y, t)$, represents the minimal time cost for the pedestrians at a given location to move a unit distance, which is determined by the density of pedestrians at this location. The potential function, written as $\phi(x, y, t)$, defines the time to reach the final destination. The potential function can be calculated given that the neighboring points with the minimal value of the cost function are always chosen from the current location to the destination. The cost function and the potential function can be quantified using an Eikonal equation:

$$\nabla \phi(x, y, t) = c(x, y, t) \qquad (11.1)$$

After the potential function is resolved from the above equation, the minimal time from any point to the destination can be obtained. The pedestrians in the crowd will choose the route with the direction opposite that of the gradient of the potential function, $\phi(x, y, t)$. Eventually, the crowd movement is governed by a conservative equation:

$$\partial \rho(x, y, t)/\partial t + \nabla(v(x, y, t)\rho(x, y, t)) = 0 \qquad (11.2)$$

11.4.2.3   Vehicle Agent Model

The vehicles running on the two one-way roads are modeled using another type of vehicle agent, with each agent representing an individual vehicle. A road consists of four lanes, and each agent only moves along one lane and will not switch to another. All vehicle agents tend to maintain the maximum speed allowed by the traffic control. A safety distance is defined to regulate vehicle agents according to their speeds. When two adjacent agents become too close, the one that is behind will attempt to slow down to maintain a safe distance between them. When an agent encounters a red traffic light, it stops to leave the crosswalk to the pedestrians. When a vehicle agent approaches the end of a road, it slows down for turning.

In summary, Table 11.4.2.3 lists the basic features of the three types of models:

TABLE 11.1
Features of Simulation Models

Models	Pedestrians	Crowd in the Assembly Area	Vehicles
Type	Agent based	Computational	Agent based
Scale	Individual	Crowd(macroscopic)	Individual
Outputs	Each pedestrian's behaviors	A crowd's dynamic distribution pattern	Each vehicle's motion pattern
Environment	2D space	2D space	1D space

11.4.3 Crowd Simulation over the Hybrid Grid Simulation Infrastructure

The overall scenario consists of two main regions, that is, the square and the surrounding roads involving entities with very different dynamics. Accordingly, two sets of simulation models were developed to represent the evolving activities in the two regions. Each set of models forms an individual simulation federation. The square federation and the roads federation operated on two administrative domains correspondingly (marked as federations A and B, respectively, in the remainder of this chapter). The former federation (referred to as federation A) simulates how aggregated pedestrians disperse from the assembly area (see Figure 11.6) and permeate over the whole square. The latter federation (referred to as federation B) describes the pedestrian flows on the crosswalks and the motion patterns of the vehicles on the roads. In particular, the combined influence from the dynamics of the crowd on the square and from traffic control on each pedestrian flow has been considered. Figure 11.8 illustrates the structure of the overall agent-based models sustained by the infrastructure. The major event or data flows among individual federates are presented.

Twenty thousand pedestrians were simulated in the scenario, and this requires the same large number of deliberative agent which inevitably results in a performance bottleneck for a desktop computer. Aiming at the potential, the square was divided into seven partitions as shown in Figure 11.9. Each dashed line represents a border between two neighboring partitions. Dark gray blocks denote the areas restricted from access. Six pedestrian federates (denoted by A1 through A6) were constructed with each corresponding to a partition linked to a crosswalk or passage to simulate the activities within the partition. A pedestrian federate consists of a number of pedestrian agents. Another crowd federate (A7) simulates the crowd initially aggregated in the assembly area using the macroscopic model. Each partition of the virtual environment was maintained by a federate, respectively, as illustrated by Figure 11.8, which highlights the interactions among federates A4, A6, and A7 (the interactions between entities inside a federate are detailed in Section 11.4.2):

- *A pedestrian federate to another.* If a pedestrian moves to an adjacent partition, the ownership of the corresponding agent will be transferred to the federate maintaining this partition. This partition-crossing activity will be signaled by an event sent from the source federate to the destination federate, which carries the information of agents states, the locations, and the time point this movement occurs.

- *The crowd federate to an adjacent pedestrian federate.* The crowd federate provides statistical information about the pattern of the crowd dispersing from the assembly area. The information drives each adjacent pedestrian to create agents at certain locations and times as well as to initialize the state of each agent. This results in a gradual de-aggregation of the crowd in the assembly area, which was originally in an ultra high density.

- *A pedestrian federate to the adjacent crowd federate.* A pedestrian may move back to the assembly area from an adjacent partition; thus the departing agent will aggregate to the crowd simulated by the crowd federate. The macroscopic model needs to adapt to this event. Federation B simulates vehicle and pedestrian activities on the roads. Each road and the vehicles running on them are simulated by an individual road federate (denoted by B1 and B2). A road federate maintains a number of vehicle agents. The activities of pedestrians on each crosswalk are simulated by an individual pedestrian federate (denoted by B3 through B6 corresponding to the crosswalks illustrated in Figures 11.6 and 11.8).

- *Between a pedestrian federate (in square) and an adjacent pedestrian federate.* Taking federate A6 and federate B6, for example, at a time point at which traffic light switches, federate B6 signals federate A6 with an event. If the light becomes green to pedestrians waiting in front of the crosswalk, federate A6 responds with an event of pedestrian moving to the crosswalk. Federate B6 will be fed with the information of the pedestrians, and federate B6 accordingly initiates a procedure of simulating agent movement on the crosswalk.

- *Between a pedestrian federate (in roads) and a related vehicle federate.* Federate B2 simulates the vehicles running on a road (B2). When the traffic light of the crosswalk associated with B6 remains green (from vehicles' perspective), federate B2 sends events to federate B6 about the vehicles crossing the crosswalk. When the traffic light switches, federate B6 signals federate B2 with an event. If the light turns to red or green, vehicle agents on federate B2 will respond to this event as described in Section 11.4.1.

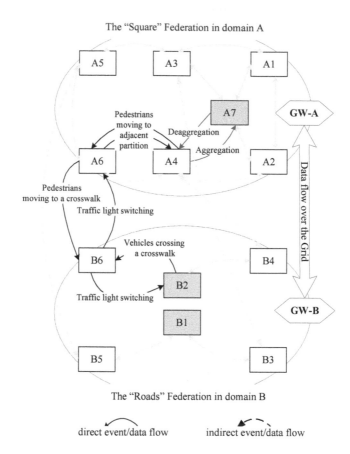

FIGURE 11.8
Structure of the crowd evacuation simulation system.

In the whole Crowd simulation, one virtual time unit in the simulation represents 0.1 second wall clock time. Sixty-minute activities in the scenario were simulated, which corresponds to 36,000 virtual time units. The pedestrian federates' timesteps and lookaheads [72] are set to 2.0, and the vehicle models' timesteps and lookaheads are set to 1.0. As each traffic light signals every 60 seconds, the lookahead of both gateways are set to 600. This value appropriately reflects the relation of the two federations. One virtual space unit represents 1.0 meter in the scenario. The resolution of the simulated square and crosswalks are set to 0.5 meters, while the resolution of the roads is set to 1.0 meter.

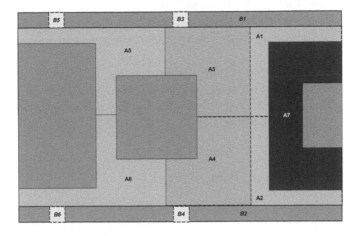

FIGURE 11.9
Conceptual view of the partitions of the scenario.

11.5 Experiments and Results

The test bed is built upon two compute centers at Indiana University, United States, and at Huazhong University of Science and Technology, China. Each site contains a compute cluster. Table 11.5 describes their resource configuration. Grid services are provided by Globus Toolkit 5.0.3 [51].

TABLE 11.2
Test Bed Description

Compute Center	Resource Description
Univ. Indiana, US	Compute cluster: 20 nodes Node: Intel® Core™ i5-2520M Processor 2.50GHz, 4G RAM Network: 1Gbps Ethernet OS: Ubuntu Linux 10.04 LTS
CUG, China	Compute cluster:10 nodes Node: Intel® Xeon® Processor E5603, 2G RAM Network: 2Gbps Myrinet OS: SuSE Linux Enterprise 11sp1

11.5.1 Communication Latency

We first performed a latency benchmark to investigate the communication overhead incurred by the underlying infrastructure. The benchmark measures the latency of federate communications as follows: One federate (on A1) sends an attribute update [25], and upon receiving this update, the other federate (on B1) sends it back to the sending federate. The elapsed time of this communication is calculated using the real-time taken at the sending and reflecting federates [16, 21]. The averaged result indicates that the latency is about 373 milliseconds.

11.5.2 Crowd Simulation Outputs

The experimental data recorded from one run were collected and the evolvement of the whole scenario is reconstructed. Snapshots of the virtual scenarios states of the two regions at different stages are presented as follows:

- *Initial Stage of the Activities in the Square.* There are in total 30,000 agents with their locations set conforming to uniform distribution in the assembly area. Those represent the state of the crowd initialized by federate A7 (the computational crowd model) according to the density function. The speed (see Section 11.4.2.2 for definition) of the crowd remains zero until the occurrence of the event *End_of_the_ceremony*.

- *Middle Stage of the Activities in the Square.* After triggering the *End_of_the_ceremony* event, agents at federates A1though A4 were initialized and driven by the inputs from A7 (see Figures 11.8 and 11.9) to represent the scattering pedestrians. With the simulation progressing, agents at models A5 and A6 were eventually driven to simulate the activities of pedestrians on partitions A5 and A6. We developed a visualization end for displaying the pedestrians in the square area, and Figure 11.10(A) presents the overall distribution of pedestrians at simulation time 9,000. The coordinates of the remaining pedestrians are exactly the agents current positions logged by A1 through A6. Agents are marked with three different colors: red (a member of a tourist group), blue (a member of a group of friends), and black (an independent individual). We selected two areas as shown in Figure 11.10(A) to have a zoom-in view. The zoom-in view explicitly gives each pedestrian's position and moving direction. Figure 11.10(B) presents a number of agents in the middle of the square in which exist two tourist groups. We can see the agents in a group are close to each other and moving toward the similar directions. Figure 11.10(C) highlights a selected area at the entrance of passage P1. It can be observed that pedestrians are swarming towards the entrance of the crosswalk.

- *Middle Stage of the Activities over a Crosswalk.* Federates B3 through B6 generate the activities of the pedestrians moving over the crosswalks. Figure 11.11 presents the snapshot of the walking pedestrians over crosswalk

P4 at simulation time 9,030. The figure also shows the status of vehicles close to P3. Each black block represents a vehicle with an arrow marking its direction and the length of the arrow indicating its rate. A block associated with no arrow denotes a still vehicle.

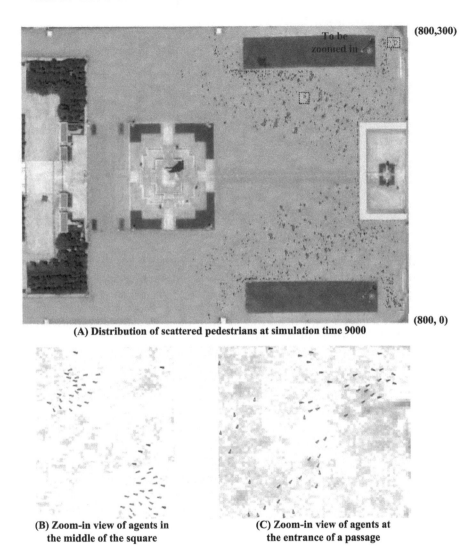

(800,300)

(800, 0)

(A) Distribution of scattered pedestrians at simulation time 9000

(B) Zoom-in view of agents in the middle of the square

(C) Zoom-in view of agents at the entrance of a passage

FIGURE 11.10
Snapshot of pedestrians in the square at simulation time 9000.

The activities on a road are presented in Figure 11.12. It gives a snapshot of the vehicles running on a road (B2) at simulation time 9,090, at which point pedestrians are restricted from trespassing the crosswalks by traffic control.

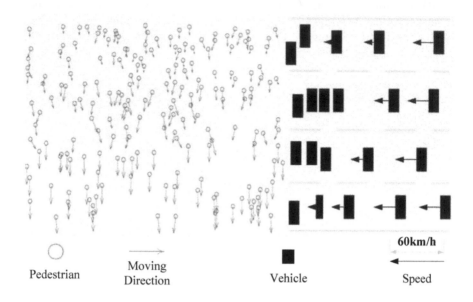

FIGURE 11.11
Snapshot of pedestrians and vehicles on a road at simulation time 9030.

We can observe that the vehicles located closer to both ends of the road have a speed lower than those in the middle.

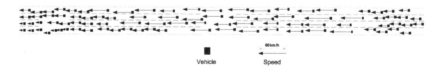

FIGURE 11.12
Snapshot of the vehicles on a road at simulation time 9090.

11.5.3   Performance Evaluation

The total averaged execution time of the simulation reports about 2118 seconds; therefore a faster-than-real-time method has been enabled to study the simulated scenario and the potential bottleneck of executing large number of deliberative agents has been successfully resolved.

In order to investigate the runtime performance of the hierarchical Grid architecture, we compare the Crowd simulation over the architecture (referred to as the Grid-enabled Crowd simulation, GCS) versus a stand-alone compute node at the CUG side (see Table 11.5 for specification). We developed a pedestrian agent model that covers the crowd activities of the square (i.e., the combination of federates A1 through A6) and uses the log of federate A7 as input to initialize the agents originating from the assembly area and the logs of federates B3 through B6 as inputs to throttle the actions of agents that finally exit the scenario via passages P3 through P6. This model is referred to as the sequential Crowd simulation, SCS, in this section.

We performed a series of experiments that focuses on the execution times of GCS and SCS. We trust that the complexity of the two simulation systems is basically controlled by the number of pedestrians, that is, the number of the deliberative agents.

Table 11.5 gives the execution times of the two types of Crowd simulation systems with increasing crowd sizes. When there are 3000 pedestrians, the execution time of GCS is 1318s which longer than that of SCE. This is due to the high communication overhead of the Grid services. With the complexity increasing significantly, GCS performs better when the communication overhead is covered by the benefits gained by distributing the computation load of agents to multiple compute nodes. A single compute node can sustain about 8000 agents as the maximum. The results indicate that (1) the hierarchical Grid architecture can significantly improve the runtime performance of executing complex simulation scenarios and (2) it scales well with the scenario's complexity.

TABLE 11.3

Performance of Sequential and Parallelized COS Systems.

Number of agents	Execution time(Sec)	
	SCS	GCS
3000	897	1318
5000	1965	1571
8000	2923	1943
10000	N/A	2080
20000	N/A	2117
30000	N/A	2518

11.6 Summary

This chapter presented a hybrid approach to modeling and simulation of a huge crowd upon a hierarchical Grid simulation architecture. We successfully developed a simulation system of the evacuation procedure of a huge crowd in an urban area that initially had a high density. Three heterogeneous models, namely a computational microscopic crowd model, a pedestrian agent model, and a vehicle agent model, have been constructed to characterize different portions and aspects of the large and complex simulated scenario.

The simulation of a huge crowd at the scales of both the individual and the crowd has been successfully developed with the support of the hierarchical Grid simulation architecture. The simulation outputs indicate that the proposed approach is effective in dealing with the size and complexity of scenarios involving huge crowds. The performance issue in connection with executing a large number of deliberative agents has been properly addressed as well. These achievements are based on the underlying simulation architecture empowered by Grid computing. Coarse-grained and fine-grained models can be seamlessly interplayed to describe the global dynamics of the whole while being able to capture the individuality of pedestrians. The experimental results also indicated that even the execution efficiency of fine-grained models can be guaranteed when the asymmetry of the simulation models in different domains is properly mapped to the infrastructure. The infrastructure well suits compute-intensive and large-scale simulation.

It is possible extend the current design to support large simulation with models using hybrid computing platforms, for example, incorporating many-core architectures [24]. It is particularly interesting to adapt multi-scale ecosystem simulations and interacting neuronal networks [84] to the infrastructure.

12

Massively Parallel Modeling & Simulation of a Large Crowd with GPGPU

CONTENTS

12.1 Introduction

A human crowd is a fascinating social phenomenon in nature. A crowd of people may show a well-organized structure and become disordered animals at another point. Numerous incidents/accidents in connection with a crowd have been recorded in human history. How to predict and control the behavior of a crowd upon various conditions/events is an intriguing question faced by many psychologists, sociologists, physicists, and computer scientists. It is also a major concern of many government agencies when dealing with crowds in confrontation.

A crowd is not simply a collection of individuals; if usually exhibits highly complex dynamics. The study of crowds in confrontation operations has received more and more attention. Crowd gatherings accompanied by severe violence have occurred frequently in our nowadays restless world. From a researcher's perspective, another important reason is that the dynamics of a crowd in a confrontation operation are largely influenced by external stimuli (properties and status of entities in the scenario as well as events), which are highly uncertain and often interact with the collective behavior of the crowd.

In general, pure mathematical approaches or analytic models are not adequate in characterizing the dynamics of a crowd. Crowd modeling and simulation (M & S) has recently been gaining tremendous momentum [121]. Existing models are largely at two extreme levels (microscopic and macroscopic): either model each individual as an autonomous agent, or treat the crowd as a whole [26, 57, 58] or consisting of homogeneous particles [51] with no cognitive features.

No matter that a crowd is formed spontaneously or organized, individuals in the crowd gathering at the same time and space will globally exhibit common features, which can be well described by macroscopic modeling approaches. But the inherent pitfall of macroscopic modeling approaches is the incapability to reflect the impact brought by regional events and individualities within the crowd: for what microscopic approaches are designed. On the other hand, in general there is lack of a formal method to formulate a crowd's common features with agent-based approaches.

The collective behavior of a crowd in a confrontation operation is determined by both unanticipated external stimuli and the common features of the crowd itself. In this study, we propose a novel method based on the concept of vector field to formulate the way in which external stimuli may affect the tendency of the behaviors of individuals. Our approach represents each individual as an autonomous agent whose actions are guided by the vector field model. As such, we bridge the gap between the macroscopic with the agent-based approaches to more accurately characterize the interaction dynamics between a crowd and external stimuli.

Furthermore, as pointed out by Helbing, emotions play a decisive role in how people behave in crowds and the more nervous crowds get, the more unpredictably and irrationally they behave. Existing crowd models normally incorporate a number of tangible factors (such as speed, location, appearance, age) and some also consider intangible (such as emotional) factors. How to properly portray intangible factors and to quantitatively measure the impact of these factors in a model remain a research challenge. This study also explores an information entropy-based method to quantify the degree of of individuals and proposes the potential for disorder of the whole crowd. A quantitative analysis on the intangible dynamics of a crowd in confrontation is then enabled. It is a research challenge to support confrontation operation simulations (COS) involving a large crowd. Large agent-based systems, such as simulation of large crowds at the individual level, have long placed it in the highly computation-intensive world. Using traditional CPU-based high performance computing technology may provide an ad hoc solution to the performance issue but this type of technology is subject to a number of limitations: heat dissemination, excessive energy consumption, high-density power, and excessive cost for associated cooling systems. There exists a pressing need for computing methods for COS that can simulate a crowd of large size while ensuring energy efficiency.

In the past few years, the modern Graphics Processing Unit (GPU) has evolved into a highly parallel, multithreaded, and many-core processor far beyond a graphics engine that substantially outpaces its CPU counterparts in dealing with computationally demanding, complex problems [8]. In this study, we have developed a parallelized Crowd simulation approach, which successfully adopts general-purpose computing on the graphics processing unit (GPGPU) to thoroughly exploit the parallelisms of the COS process. The proposed approach has been developed based upon NVIDIAs Compute Unified Device Architecture (CUDA) [9], a general-purpose parallel computing architecture. Results demonstrate that GPGPU-aided approaches are remarkably superior to the distributed computing-based counterparts in terms of both performance and energy consumption.

The remainder of this chapter is organized as follows: Section 12.2 provides some background knowledge and redefines notations borrowed from other disciplines. Section 12.3 introduces the vector field approach. Section 12.4 presents a case study of the simulation of a crowd in confrontation operation. This section also gives a quantitative analysis of the evolution of the simulated crowds entropy in terms of degree of panic. Section 12.5 introduces the approach to M & S of a large crowd in a confrontation operation using GPGPU. We conclude this chapter with a summary and future work in Section 12.6.

12.2 Background and Notation

This study (1) adopts an agent-based approach for modeling individuals, (2) transplants the concept of vector field to reflect the influences of external stimuli on a crowd, (3) uses information entropy to analyze a crowds intangible dynamics, and (4) adopts GPGPU to parallelize COS to sustain scenarios consisting of large crowds. Several important notations related to the above methods are presented as follows:

- The *agent-based approach* is currently the most active approach used for crowd M & S in the community of computer science and engineering [15, 24, 51, 23, 111, 112]. A crowd is regarded as a collection of heterogeneous individuals who are empowered with decision-making capability, with each agent representing an individual. The agent-based approach is the most natural way to model behaviors with strong individual differentiations. A typical example is the PAX system, which provides an M & S tool for scenarios of peace support missions [98]. Our model also uses autonomous agents to model individuals.

- A *vector field* is a construction in vector calculus that associates a vector to every point in a subset of Euclidean space. Vector fields are often used in physics to model the strength and direction of some force, such as the magnetic or gravitational force, as it changes from point to point. A vector field in the context of this study formulates the relationship between people's intention (internal) and the external stimuli. The vector field theory in physics will be adopted as the mathematical basis of the macroscopic model. A vector field (see Figure 12.1) describes the influence (force) that a stimulus has on a certain intention of a person who perceives this stimulus in the scenario's space. An intention is regarded as a *charged particle*, whose *charge* is subject to the intention's *magnitude* and may change with the evolution of simulation. A vector field is *dynamic* if its intensity and direction can change over time; otherwise it is *fixed*.

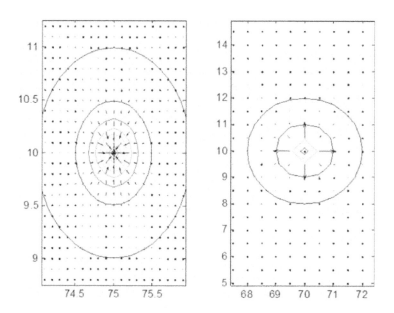

FIGURE 12.1
A gravitational field (left) and a repulsive field (right).

In this study, a *vector field* is defined for *one particular stimulus* that merely works on *one particular intention*. As such, interference does not exist between any two vector fields. This definition significantly differs from those vector fields in physics. The effect of multiple vector fields on an individual is only exhibited by the combined force resulted from the forces these vector fields act on multiple intentions exclusively.

- *Entropy* has important physical implications as the degree of disorder of a physical system [54, 75, 107]. Information entropy is a measure of the uncertainty associated with a random variable (X), which is defined as

$$H(X) = H(P_1, P_2, \ldots, P_n) = -\sum P(x_i) \log P(x_i) \qquad (12.1)$$

where $P(x_i)$ is the probability that X is in the state x_i, and $\sum_n P(x_i) = 1$.
If $P(x_i) = 0, P(x_i) \log P(x_i)$ is defined as 0. The more disorderly a system
is, the more information it contains, and vice versa. In the context of our
crowd model, five types of behaviors have been defined, namely following,
avoiding, adjusting, confrontation, and retreat. An agent may be specified
a behavior at any point to conduct subject to a probability distribution on
these candidate behaviors. The information entropy can then be calculated
for the whole crowd. The value of the entropy will provide a quantitative
measure on how disorderly the current crowd is, which may then facilitate
controlling the crowd.

• The notation *Degree of Parallelism (DoP)* is used to quantify the parallelism as a problem to be solved. A problems DoP means the number of
portions in the problem, which can be concurrently solved/executed with
the same results as those attained in a serial manner.

12.3   Hybrid

A hybrid behavior model is designed to manipulate each agent's behavior.
Figure 12.2 presents a conceptual view of the proposed model consisting of
two submodels, that is, (1) a rule-based submodel to specify each agent's exact
behavior and (2) a vector field submodel representing the influences of external
stimuli common among all agents in the simulation. We adopted a classic
design for an agent, which has the cognition capability to sense, deliberate,
and then act. Rule-based agent approach has been extensively covered by
existing work. This study emphasizes the vector field submodel only.

The vector field submodel maintains a set of vector fields. The submodel
computes the integrative influence of the external stimuli on people's various
internal intentions, and it outputs the *tendency* (measured by the combined
force) of an individual's behavior. The tendency means what the individual
is likely to do rather than a *deterministic* action/motion as in Helbing's approach [51]. The vector field approach has been examined in a scenario of
demonstration in front of a governmental building, as shown at the top of
Figure 12.3.

A crowd of demonstrators moves on a march toward a governmental building (with its entrance highlighted as the red star). Each individual demonstrator is represented by a small red circle with its field of view and mov-
ing/confrontation direction marked. Armed policemen are denoted by blue

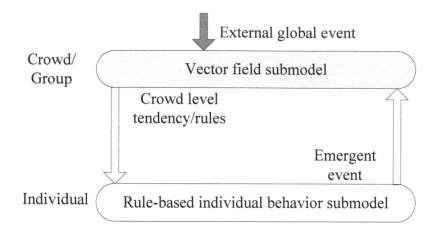

FIGURE 12.2
Conceptual view of the vector field submodel and the behavior model.

triangles pointing at their moving/confrontation directions. Agents are moving within the 2D space confined by the upper and lower bounds.

Here we consider an agent's intentions of two types: (A) to approach the entrance of the governmental building; and (B) to avoid being attacked by the policemen. A basic type of vector fields has been defined to represent the influences of the governmental building on an agent's intention A ($\vec{E}_g(\vec{r})$ in Equation (12.2)) and (2) each armed policeman on an agent's intention B ($\vec{E}_{p[i]}(\vec{r})$ in Equation 4):

$$\vec{E}_g(\vec{r}) = \begin{cases} C_A \dfrac{(\vec{r} - R_g)}{\left|\vec{r} - R_g\right|} & \left|\vec{r} - R_g\right| < D_1 \\[2em] k_A \dfrac{Q_g(\vec{r} - R_g)}{\left|\vec{r} - R_g\right|^3} & D_1 \leqslant \left|\vec{r} - R_g\right| \leqslant D_{MAX} \\[2em] 0 & \left|\vec{r} - R_g\right| \geqslant D_{MAX} \end{cases} \tag{12.2}$$

$$\vec{E}_{p[i]}(\vec{r}) = \begin{cases} -k_B \dfrac{Q_{p[i]}(t)(\vec{r} - R_p)}{\left|\vec{r} - R_p\right|^3} & \left|\vec{r} - R_{p[i]}\right| < D_2 \\[2em] 0 & \left|\vec{r} - R_{p[i]}\right| \geqslant D_3 \end{cases} \tag{12.3}$$

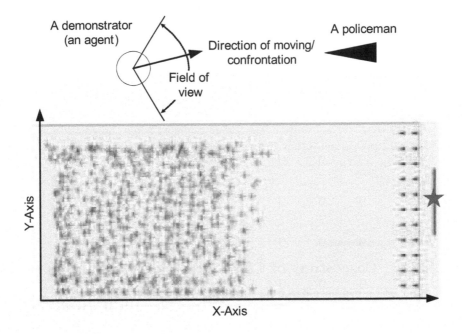

FIGURE 12.3
Simulation of a crowd in a confrontation operation.

where Q_g is a fixed variable representing the intensity of the governmental building; $Q_p(t)$ represents the intensity of the governmental building; $\vec{r}$ is a vector from the origin of coordinate to an agents location; R_G is the vector (fixed) from the origin to the star; R_G is the vector from the origin to a policeman (i); and C_A is a constant. D_1, D_2, and D_3 ($>> D_2$) are constants representing distances; D_{MAX} represents the max distance between any two locations in the scenario; and k_A and k_B are two coefficients.

We write an agent's intentions A, B at a time point as $Q_A(t)$ and $Q_B(t)$ respectively. The combined effect of the governmental building and the policemen ($i = 1, 2, \ldots, n$) on the agent can be written as

$$\vec{F}_{AB}(t) = \vec{F}_A(t) + \vec{F}_B(t) = \vec{E}_g(\vec{r}) \times Q_A(t) + \sum_{i-1}^{n} \vec{E}_{p[i]}(\vec{r}) \times Q_B(t) \qquad (12.4)$$

In this scenario, policemen are deployed between the governmental building and the demonstrators, and it is close to the government. Given that only intentions A and B are concerned, the agent tends to confront the policemen when $\left|\vec{F}_{AB}(t)\right|$ is small enough (ε). The effect of $\vec{F}_{AB}(t)$ on the agent relies on the component (A-component) of $\vec{F}_{AB}(t)$ along the direction of $\vec{F}_A(t)$; negative means in the same direction (see Figure 12.4). Let α denote the angle between $\vec{F}_{AB}(t)$ and $\vec{F}_A(t)$; the magnitude of the A-component of $\vec{F}_{AB}(t)$ is written as $\vec{F}_\varepsilon(t)$:

$$\vec{F}_\varepsilon(t) = \vec{F}_{AB}(t) \times \cos(\alpha) \tag{12.5}$$

When $\vec{F}_\varepsilon(t)$ is negative, the agent tends to leave the governmental building and policemen; otherwise the agent tends to approach the governmental building. The intensity of the tendency of the agent's behavior is proportional to $\left|\vec{F}_\varepsilon(t)\right|$.

12.4 Case Study of Confrontation Operation Simulation

Simulation has been executed using the agent model based on the vector-field method to examine the effectiveness of the proposed method. The dynamics of the simulated system have been quantified via entropy calculation afterward.

12.4.1 Simulation of a Crowd in a Confrontation Operation

The simulation scenario involves a crowd of 500 demonstrators marching to the governmental building and interacting with the policemen (22 on initialization) attempting to expel the demonstrators. The simulation lasts for 200 time units.

Figure 12.3 (see description in Section 12.3) illustrates the initial stage of the simulation. The gravity field, $\vec{E}_g(\vec{r})$, dominates the agents' (marked in red) tendency; $\vec{F}_\varepsilon(t)$ on the agents are positive, so the agents approach the red star. A small number of agents in the front lead the way, followed by the remaining agents.

Figure 12.5 demonstrates three other stages of the simulation. When the agents in the front get close enough to the policemen, the, $\vec{E}_{p[i]}(1\ n)$, generate a repulsive force sufficient to balance $\vec{E}_g(\vec{r})$. $\vec{F}_\varepsilon(t)$ on some agents in the front half become less than ε, thus most of them confront the policemen and their color turns to green (see Figure 12.5(A)).

Policemen start to move toward the agents to prevent the demonstrator agents from further approaching the red star (see Figure 12.5(B) and (C)). When the distance between the policemen and the agents in the front is generally small, the agents panic level increase, and the intensity of their intention A diminishes. In these cases $\vec{E}_{p[i]}(1\ n)$ dominate, and $\vec{F}_\varepsilon(t)$ of most agents become negative which drive more and more agents leave the demonstration.

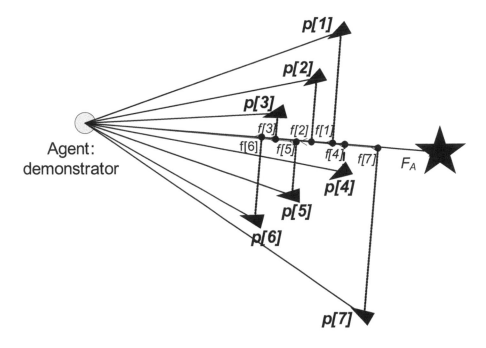

FIGURE 12.4
Combined force acted on an agent.

12.4.2 Dynamics Analysis via Entropy Calculation

In this study, we introduce the concept of information entropy to analyze the degree of disorder of the simulated crowd. The definition of entropy is available in Section 12.2.

Figure 12.6(A) presents the information entropy of the crowd calculated for the simulation scenario presented in Section 12.4.1:

- A peak in entropy in the duration, from time 0 to 40 can be observed. This is caused by the diversity of agents' behaviors when the agents pour into the demonstration at the beginning.

- After time 20, the crowd march to the governmental building and most agents' behaviors converge; thus the value of entropy drops sharply.

- From time 40 to 70, there is a sharp increase in information entropy. Agents are in the process of approaching the governmental building, and their degrees of panic are growing as they get closer to policemen.

- After time 70, agents continue to move forward and most of their behaviors become confrontational to the policemen. This results in a drop in the value of information entropy.

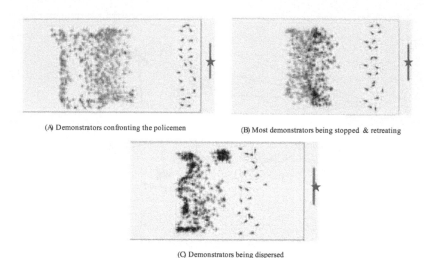

(A) Demonstrators confronting the policemen

(B) Most demonstrators being stopped & retreating

(C) Demonstrators being dispersed

FIGURE 12.5
Evolvement of a simulated crowd in a confrontation operation.

• From time 100 to 130, the policemen move toward the agents in advancing. Agents gradually start to retreat. As the number of retreating agents increases, the behaviors of the crowd tend to converge, and the entropys value drops further.

• When some small groups of agents are formed and remain in confrontation with the policemen, the third peak is reached at time 150. There can be few agents who managed to cross the policemen and rush into the governmental building. This diversity makes the information entropy at a relatively high level.

In the previous simulation, agents with strong intention A may break through the policemen line. To test the influence of more policemen on the order of the crowd, we performed another simulation with more policemen (30) deployed when most agents are in the state of confrontation. From Figure 12.6(b), we can see that the value of information entropy in the current simulation is generally less than that calculated from the previous simulation. The information entropy also drops to zero at the end stage of the simulation, in contrast to the results presented in Figure 12.6(a). This denotes that the crowd is more orderly in this simulation and the probability of an unanticipated emergency events is reduced. The time for the crowd to reach stability is also shortened. In the two simulations, the information entropy properly

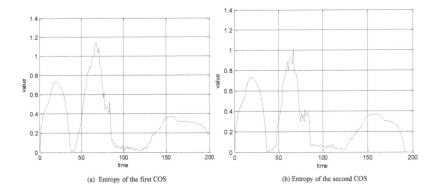

(a) Entropy of the first COS   (b) Entropy of the second COS

FIGURE 12.6
The information entropy of the case studies of COS.

reflects the status and the evolution of the dynamics of the crowd under the influence of the policemen and the governmental building.

12.5 Aided by GPGPU

Due to the complexity of the COS at the individual agent level, the size of a simulated crowd is very limited, that is, around 8000 individuals as the maximum, even using a cutting-edge desktop computer. Another problem is the execution efficiency. We developed a GPGPU-aided solution to address these problems.

12.5.1 Parallelization of Crowd Simulation

The most intensive computation of a COS in this study lies in the execution of agents. The complexity increases almost linearly with the size of the crowd, that is, the number of agents. A sequential COS operates in a number of identical virtual time frames, typically representing 0.2 seconds in real world. For each time frame, the simulation executes all agents one by one to compute each agents velocity (V), position (P), decision on behavior selection (B), and its target (G). Computing the four attributes of an agent requires the COS system state (obtained from the last time frame), and this is independent of the results of any other agent in the current time frame.

This means that it is possible to parallelize the task of executing multiple agents. Considering the feature of the Crowd simulation, we propose a scheme to partition the execution of agents into a number of subtasks, with each executing an individual agent as shown in Figure 12.7. Given that a COS scenario consists of NA individuals, the COS' DoP equals NA at the individual agent level.

12.5.2   Evaluation of Performance and Energy Efficiency

We have performed a number of experiments to study the performance of the alternative COS systems. The configurations of the test bed are presented in Table 12.1. The original sequential version of the COS was modeled using Java upon RePast 3.0 [16], a Java-based toolkit for the development of lightweight agents and agent models. RePast has become a popular and influential toolkit, providing the development platform for several large multi-agent simulation experiments, particularly in the field of social phenomena. We first examined the overhead distribution of the simulation program. The computer node is only able to execute 8000 demonstrators to the maximum. Given a scenario consisting of 4000 demonstrators with the timestep set as 0.5 seconds (simulation time), the simulation execution time on a single computer node is 309 seconds. The elapsed time in calculating V, P, B, and G is 308 seconds, which contributes about 99.6% of the overall overhead. Clearly, the performance bottleneck of the program lies with calculation of the four attributes of the agents.

12.5.2.1   GPGPU-Aided Confrontation Operation Simulation

Based on the above observations, we developed a parallelized simulation using GPGPU, which excels in handling large numbers of concurrent fine-grained subtasks. The GPGPU-aided COS uses an individual CUDA thread to compute each agent's four attributes in each time frame (see Figure 12.7). The new simulation program was developed based on JCuda (version 0.3.2a) [54], a Java bindings for Java programs to interact with CUDA runtime and driver APIs. Thus, most of the original simulation code in Java can be reused while still having the benefit from the underlying powerful parallel programming and computing capabilities offered by GPGPU.

Given that a COS scenario consists of NA individuals, we produced a scheme that maps this task to CUDA threads in the following steps: *Step 1*: After initialization of the kth time frame, the host assigns NA data sets derived from the simulation's current system state. Each data set corresponds to an individual agent for computing its velocity, position, decision on behavior selection, and its target; *Step 2*: The host invokes NA CUDA threads via JCuda, and these threads are evenly grouped into 480 blocks (NA/480 threads operating in each block) on the device. This means that 480 cores on the GPU are assigned to execute these agents with each core computing one agents attributes individually; *Step 3*: Step 2 repeats until the NA threads complete execution. The NA agents' new attributes are then passed from the device

TABLE 12.1

Configuration of the Test Bed.

Specifications	Computers	
	Desktop Computer (Client)	Computer Cluster (Server, 1 master node and 15 worker nodes interlinked via 1Gbps Ethernet)
Operating System	Windows 7 Professional	Rocks 4.2.1 (Cydonia) Red Hat Enterprise Linux ES release 4, X86_64 2.6.9-42.ELsmp
CPU	Intel Pentium Dual Core at 3.20GHz and 3.19GHz	2 x Intel Dual Core at 3.0GHz
RAM	2048M	4096M
Power Consumption	650W	3.5KW~14KW
Specifications of NVIDIA GeForce GTX 480		
CUDA Cores	480	
Processor Clock	1401MHz	
Standard Memory	1536 MB GDDR5	
Memory Bandwidth	223.8GB/s	
Power Consumption	250W	

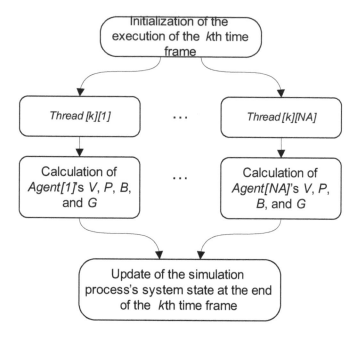

FIGURE 12.7
Execution task graph of a parallelized simulation scheme.

to the host. The host then updates the system state through the RePast simulation engine and enters the $(k+1)$th time frame. Our design minimizes the thread number in a thread block while it creates thread blocks as many as possible when executing threads of this type. Hence, these threads can make the most of CUDA cores to deal with intensive computations and occupy as much fast shared memory (manipulated by each block) as possible to buffer the intermediate data generated during their executions. The GPGPU-aided COS can support scenarios consisting of more than 30,000 demonstrators.

12.5.2.2   Performance Evaluation and Energy Efficiency Analysis

In order to investigate the potentials of traditional CPU-based high-performance techniques, for example, cluster computing, and to establish a reference for evaluating the GPGPU-based approach, we developed another parallelized COS with the support of HLA_RePast [16], a middleware that supports the execution of multiple interacting instances of RePast agent-based models. Thus a Cluster-aided COS (CCOS) has been established, and the load of executing the original COS can then be distributed over the fifteen worker nodes of the computer cluster (see Table 12.1).

We performed a series of experiments that focused on speedup (comparing to sequential COS, referred to as SCOS) and aimed to investigate and compare the performance of GPGPU-aided and Cluster-aided COS systems. Table 12.2 gives the execution times of the two types of COS systems with different numbers of agents (demonstrators). The results indicate that (1) the two parallelized COS systems significantly improve the runtime performance and scale well with the number of agents; and (2) the GPGPU-aided COS (referred to as GCOS) always excels in performance improvement. The results (agent number $\leqslant$ 8000) are highlighted in Figure 12.8.

TABLE 12.2
Performance of Sequential and Parallelized COS Systems

Number of Agents	Sequential COS	Cluster-Aided COS (parallelized)		GPGPU-Aided COS (massively parallelized)	
	Execution Time (sec)	Execution Time (sec)	Speedup	Execution Time (sec)	Speedup
3000	206	102	2	20	9.8
4000	309	138	2.2	31	10
5000	565	191	3	57	10
8000	1219	439	2.8	95	13
10000	N/A	680	N/A	115	N/A
12000	N/A	718	N/A	129	N/A
20000	N/A	1409	N/A	209	N/A
30000	N/A	N/A	N/A	317	N/A

The GPGPU-aided COS in this study operates on a graphic card with power consumption that amounts to the maximum 250W included in the maxim 650W power consumption of the desktop. During the execution of GCOS on the desktop (viewed as a CPU-GPU hybrid system), the desktop's consumption was about ~210W. In contrast, during the execution of SCOS on the desktop (viewed as a pure CPU system), the desktop's consumption was about ~130W. The computer cluster's power consumption amounts to ~9KW during the execution of CCOS.

Taking the test scenario with 8000 agents for example, the execution times with SCOS, CCOS, and GCOS are 1219s, 439s, and 95s, respectively. The total energy consumption using the three systems is ~158470j, ~3951000j, and ~19950j. Comparing with the GCOS system, SCOS/CCOS consumes ~7/~197 times more energy. This analysis does not consider the power consumption of the cooling system for the computer cluster room. The experimental results demonstrate the great advantages of GCOS over SCOS and CCOS in terms of both runtime performance and energy consumption.

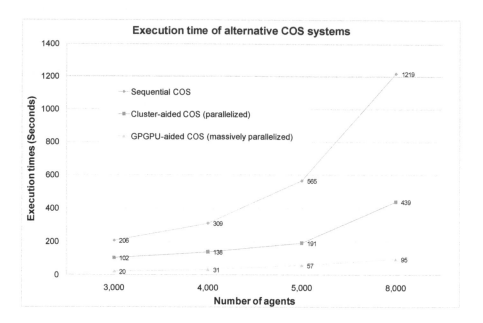

FIGURE 12.8
Execution time of alternative COS systems (agent number ⩽ 8000).

12.6   Summary

This study explored an energy-efficient and high-performance solution to simulation of confrontation operations involving large crowds. The novel simulation approach, namely GPGPU-aided COS, was developed to address the scalability and performance issues using GPGPU.

We first proposed a vector field method that aims to formulate the way in which external stimuli may affect the tendency of the behaviors of individuals. Together with the agent-based approach, a model for the simulation of a crowd in confrontation operations has been established using RePast. We also introduced the concept of information entropy to analyze how the change in each individual's behavior may affect the intangible dynamics of the whole crowd. A case study of Crowd simulation was carried out. Through the measurement of information entropy, the status and the evolution of the dynamics of the crowd can be revealed. The results indicate that (1) the proposed COS model can exhibit the typical behavior pattern of a crowd in confrontation; and (2) that information entropy can provide evident support for the design of control tactics for crowd control.

This study then emphasizes the feasibility and effectiveness of COS with

GPGPU. The GPGPU-aided approach naturally divides a COS into a large number of fine-grained tasks; thus it effectively exploits the parallelism of the COS system at the individual agent level. It seamlessly maps the tasks to the same number of CUDA threads that can be executed concurrently by hundreds of GPU cores.

Experiments have been carried out to evaluate the performance of GPU-aided COS and to investigate the potentials of traditional CPU-based high-performance techniques. A Cluster-aided COS has been developed based on HLA-RePast. Although Cluster-aided COS runs over a high-end CPU-based computer cluster, GPU-aided COS prevails in runtime efficiency: G-EEMD is ~6 times faster than the best distributed counterpart does. More importantly, the graphics card has maximum power consumption at only ~1/36 of the computer cluster's power consumption. This figure does not consider the power consumption of the cooling system to ensure that the computer cluster is operable. The results indicate that GPGPU is a very promising technique in the simulation of social phenomena. The proposed GPGPU-aided COS is indeed a highly energy-efficient and an ultra high-performance solution to the M & S of confrontation operations

References

[1] *RTI 1.3-Next Generation Programmers Guide Version 5.* DoD, DMSO, Feb. 2002.

[2] Pitch Kunskapsutveckling AB. 2004.

[3] S. AbouRizk and Y. Mohamed. Simphony-an integrated environment for construction simulation. In *Proceedings of the 2000 Winter Simulation Conference*, pages 1970–1974, Orlando, Florida, USA, Dec. 2000.

[4] R.G. Addie. Quantum simulation: Rare event simulation by means of cloning and thinning. *European Transactions on Telecommunications*, 13(4):387–397, Jul./Aug. 2002.

[5] A. Agarwal and M. Hybinette. Merging parallel simulation programs. In *Proceedings of the 19th ACM/IEEE/SCS Workshop on Principles of Advanced and Distributed Simulation*, pages 227–233, Monterey, California, USA, June. 2005.

[6] C. Berchtold and M. Hezel. An architecture for fault-tolerant HLA-based simulation. In *Proceedings of the 15th International European Simulation Multi-Conference (ESM) 2001*, pages 616–620, Prague, Czech Republic, June. 2001.

[7] K.P. Birman. *Building Secure and Reliable Network Applications.* Prentice Hall and Manning Publishing Company, New Jersey, USA, 1997.

[8] C. Bouwens, D. Hurrell, and D. Shen. Implementing ownership management services with a bridge federate. In *Proceedings of the 1998 Spring Simulation Interoperability Workshop*, Orlando, Florida, USA, Mar. 1998.

[9] W. Cai, G. Li, S.J. Turner, B.S. Lee, and L. Liu. Automatic construction of hierarchical federation architectures. In *Proceedings of the* 6th *IEEE International Symposium on Distributed Simulation and Real Time Applications (DSRT 2002)*, pages 50–57, Fort Worth, Texas, USA, Oct. 2002.

[10] W. Cai, G. Li, S.J. Turner, B.S. Lee, and L. Liu. Implementation of federation management services over federation community networks. In *Proceedings of the* 17th *workshop on Parallel and Distributed Simulation (PADS 2003)*, pages 50–57, Diego, California, USA, Jun. 2003.

[11] W. Cai, S.J. Turner, and B.P. Gan. Hierarchical federations: An architecture for information hiding. In 15th *Workshop on Parallel and Distributed Simulation (PADS '01)*, pages 67–74, Lake Arrowhead, California, USA, May 2001.

[12] W. Cai, Z. Yuan, M.Y.H. Low, and S.J. Turner. Federate migration in HLA-based distributed simulation. *Future Generation Computer System*, 21(1):87–95, 2005.

[13] D. Chen and C. Bian. Towards hybrid grid infrastructure for large simulations. In *Proceedings of the 1st IEEE International Workshop on Advances of CyberInfrastructure (IWACI 2009) in conjunction with the 15th Int'l Conf. on Parallel and Distributed Systems (ICPADS'09)*, Shenzhen, China, Dec. 2009.

[14] D. Chen, B.S. Lee, W. Cai, and S.J. Turner. Design and development of a cluster gateway for cluster-based HLA distributed virtual simulation environments. In *Proceedings of the 36th Annual Simulation Symposium (IEEE Computer Society)*, pages 193–200, Orlando, Florida, USA, Apr. 2003.

[15] D. Chen, D. Li, M. Xiong, H. Bao, and X Li. GPGPU-aided ensemble empirical mode decomposition for electroencephalogram analysis during anaesthesia. *IEEE Transactions on Information Technology in BioMedicine*, 14(6):1417–1427, 2010.

[16] D. Chen, G.K. Theodoropoulos, S.J. Turner, W. Cai, R. Minson, and Y. Zhang. Large scale agent-based simulation on the grid. *Future Generation Computer Systems*, 24(7):658–671, 2008.

[17] D. Chen, S.J. Turner, and W. Cai. A framework for robust HLA-based distributed simulation. In *Proceedings of the 20th ACM/IEEE/SCS Workshop on Principles of Advanced and Distributed Simulation (PADS 2006)*, pages 183–192, Singapore, May. 2006.

[18] D. Chen, S.J. Turner, and W Cai. Towards fault-tolerant HLA-based distributed simulations. *Simulation: Transactions of the Society for Modeling and Simulation International*, 84(10/11):493–509, 2008.

[19] D. Chen, S.J. Turner, W. Cai, B.P. Gan, and M.Y.H. Low. Algorithms for HLA-based distributed simulation cloning. *ACM Transactions on Modeling and Computer Simulation*, 15(4):316–345, 2005.

[20] D. Chen, S.J. Turner, W. Cai, G.K. Theodoropoulos, M. Xiong, and M. Lees. Synchronization in federation community networks. *Journal of Parallel and Distributed Computing (Elsevier)*, 70(2):144–159, 2010.

[21] D. Chen, S.J. Turner, W. Cai, and M. Xiong. A decoupled federate architecture for high level architecture-based distributed simulation. *Journal of Parallel and Distributed Computing*, 68(11):1487–1503, 2008.

[22] D. Chen, S.J. Turner, B.P. Gan, W. Cai, and J. Wei. A decoupled federate architecture for distributed simulation cloning. In *Proceedings of the 15*th *European Simulation Symposium*, pages 131–140, Delft, the Netherlands, Oct. 2003.

[23] D. Chen, L. Wang, G. Ouyang, and X Li. Massively parallel neural signal processing on a many-core platform. *IEEE Computing in Science and Engineering*, 2011.

[24] D. Chen, L. Wang, M. Tian, J. Tian, S. Wang, C. Bian, and X. Li. Massively parallel modelling & simulation of large crowd with GPGPU. *The Journal of Supercomputing*, 2011.

[25] D. Chen, Lizhe Wang, Congcong Bian, and Xuguagn Zhang. A grid infrastructure for hybrid simulations. *International Journal of Computer Systems Science & Engineering*, 3:197–206, May. 2011.

[26] Stephen Chenney. Flow tiles. In *In Proceedings of the 2004 ACM SIG-GRAPH/Eurographics symposium on computer animation*, pages 233–242, 2004.

[27] B. Ciciani, D.M. Dias, and P.S. Yu. Analysis of replication in distributed database systems. *IEEE Transaction on Knowledge and Data Engineering*, 2(2):247–261, June 1990.

[28] DIS Steering Committee. The DIS vision, a map to the future of distributed simulation. In *Technical Report IST-SP-94-01, Institute for Simulation and Training*, Orlando, Florida, USA, 1994.

[29] K.D. Cooper, M.W. Hall, and K. Kennedy. A methodology for procedure cloning. *Computer Languages*, 19(2):105–118, Apr. 1993.

[30] A. Cramp and J.P. Best. Time management in hierarchical federation communities. In *Proceedings of the 2002 Fall Simulation Interoperability Workshop*, Orlando, Florida, USA, 2002.

[31] F. Cristian. Understanding fault-tolerant distributed systems. *Communications of the ACM*, 34(2):57–78, 1991.

[32] S.J. Cunning, S. Schulz, and J.W. Rozenblit. An embedded system's design verification using object-oriented simulation techniques. *Simulation*, 72(4):238–249, Apr. 1999.

[33] J.S. Dahmann, F. Kuhl, and R. Weatherly. Standards for simulation: As simple as possible but not simpler, the high level architecture for simulation. *Simulation*, 71(6):378–387, Dec. 1998.

[34] Om.P. Danami and V.K. Garg. Fault-tolerant distributed simulation. In *Proceedings of the 12th Workshop of Parallel and Distributed Simulation*, pages 38–45, Banff, Alberta, Canada, May 1998.

[35] S. Das, R. Fujimoto, K. Panesar, D. Allison, and M. Hybinette. GTW: A time warp system for shared memory multiprocessors. In *Proceedings of the 1994 Winter Simulation Conference*, pages 1332–1339, Lake Buena Vista, Florida, USA, Dec. 1994.

[36] M. Eklöf, R. Ayani, and F. Moradi. Evaluation of a fault-tolerance mechanism for HLA-based distributed simulations. In *Proceedings of the 20th ACM/IEEE/SCS Workshop on Principles of Advanced and Distributed Simulation(PADS 2006)*, pages 175–182, Singapore, May 2006.

[37] M. Eklöf, F. Moradi, and R. Ayani. A framework for fault-tolerance in HLA-based distributed simulations. In *Proceedings of the 2005 Winter Simulation Conference*, pages 1182–1189, Orlando, Florida, USA, Dec. 2005.

[38] E.N. Elnozahy. *Fault Tolerance in Distributed systems Using Rollback-Recovery and Process Replication*. PhD thesis, Rice University, Texas, USA, 1993.

[39] I. Foster, C. Kesselman, and S. Tuecke. The anatomy of the grid. *International Journal of High Performance Computing Applications*, 15(3):200–222, Fall 2001.

[40] G. Fox, A. Ho, S. Pallickara, M. Pierce, and W. Wu. Grids for the gig and real time simulation. In *Proceedings of the 9th IEEE International Symposium on Distributed Simulation and Real Time Applications (DSRT 2005)*, pages 129–138, Montreal, Canada, Oct. 2005.

[41] R.M. Fujimoto. *Parallel Distributed Simulation Systems*. New York: Wiley, 2000.

[42] R.M. Fujimoto and R. Weatherly. Time management in the dod high level architecture. *ACM SIGSIM Simulation Digest Archive*, 26(1):60–67, 1996.

[43] B.P. Gan, D. Chen, N. Julka, S.J. Turner, and W. Cai. Benchmarking alternative topologies for multi-level federations. In *Proceedings of the 2003 Spring Simulation Interoperability Workshop*, pages 129–138, Orlando, Florida, USA, Mar. 2003.

[44] B.P. Gan, L. Liu, S. Jain, S.J. Turner, W. Cai, and W. Hsu. Distributed supply chain simulation across enterprise boundaries. In *Proceedings of the 2000 Winter Simulation Conference*, pages 1245–1251, Orlando, Florida, USA, Dec. 2000.

[45] J.B. Gilmer, Jr., and F.J. Sullivan. Alternative implementations of multitrajectory simulation. In *Proceedings of the 1998 Winter Simulation Conference*, pages 865–872, Piscataway, New Jersey, USA, Dec. 1998.

[46] J.B. Gilmer, Jr., and F.J. Sullivan. Multi-trajectory simulation performance for varying scenario size. In *Proceedings of the 1999 Winter Simulation Conference*, pages 1137–1146, Phoenix, Arizona, USA, Dec. 1999.

[47] J.B. Gilmer, Jr., and F.J. Sullivan. Recursive simulation to aid models of decision making. In *Proceedings of the 2000 Winter Simulation Conference*, pages 958–963, Orlando, Florida, USA, Dec. 2000.

[48] P. Glasserman, P. Heidelberger, P. Shahabuddin, and T. Zajic. Splitting for rare event simulation: Analysis of simple cases. In *Proceedings of the 1996 Winter Simulation Conference*, pages 302–308, Coronado, California, USA, Dec. 1996.

[49] Globus. 2004.

[50] C. Görg, E. Lamers, O. Fuß, and P. Heegaardy. Rare event simulation. In *Proceedings of the European COST (256) Telecommunication Symposium*, volume 5807/2009, pages 365–396. Kluwer Academic Press, 2001.

[51] Dirk Helbing, I. Farkas, and T. Vicsek. Simulating dynamical features of escape panic. *Letters to Nature*, 407:487–490, 2000.

[52] J.O. Henriksen. An introduction to slx^{TM}. In *Proceedings of the 1997 Winter Simulation Conference*, pages 559–566, Atlanta, Georgia, USA, Dec. 1997.

[53] J.O. Henriksen. Stretching the boundaries of simulation software. In *Proceedings of the 1998 Winter Simulation Conference*, pages 227–234, Washington, DC, USA, Dec. 1998.

[54] Carsten Herrmann-Pillath. Entropy, function and evolution: Naturalizing peircian semiosis. *Entropy*, 12(2):197–242, 2010.

[55] http://www.cc.gatech.edu/computing/pads/fdk.html. 2004.

[56] http://www.cc.gatech.edu/computing/pads/tech highperf.html. 2004.

[57] Ling Huang, S.C. Wong, Mengping Zhang, Chi-Wang Shu, and W.H.K Lam. Revisiting Hughes' dynamic continuum model for pedestrian flow and the development of an efficient solution algorithm. *Transportation Research Part B*, 43(1):127–141, 2009.

[58] Roger L. Hughes. A continuum theory for the flow of pedestrians. *Transportation Research Part B*, 36(6):507–535, 2002.

[59] M. Hybinette. Just-in-time cloning. In *Proceedings of the Eighteenth Workshop on Parallel and Distributed Simulation*, pages 45–51, Kufstein, Austria, May. 2004.

[60] M. Hybinette and R.M. Fujimoto. Cloning: A novel method for interactive parallel simulation. In *Proceedings of the 1997 Winter Simulation Conference*, pages 444–451, Atlanta, Georgia, USA, Dec. 1997.

[61] M. Hybinette and R.M. Fujimoto. Cloning parallel simulations. *ACM Transactions on Modeling and Computer Simulation (TOMACS)*, 11(1):378–407, Jan. 2001.

[62] M. Hybinette and R.M. Fujimoto. Scalability of parallel simulation cloning. In *Proceedings of the 35th Annual Simulation Symposium*, San Diego, California, USA, Apr. 2002.

[63] M. Hyett and R. Wuerfel. Implementation of the data distribution management services in the RTI-NG. In *Proceedings of 2002 Spring Simulation Interoperability Workshop*, Orlando, Florida, USA, Mar. 2002.

[64] IEEE 1278. *IEEE Standard for Distributed Interactive Simulation*. IEEE, 1993.

[65] IEEE 1516. *IEEE Standard for High Level Architecture*. IEEE, 2000.

[66] IEEE 1516. *IEEE Standard for High Level Architecture*. IEEE, 2001.

[67] IEEE 1516.3. *IEEE Recommended Practice for High Level Architecture(HLA) Federation Development and Execution Process(FEDEP)*. IEEE, 2003.

[68] Electronic Arts Inc. In *Westwood Studio*, 2004.

[69] K. Iskra, G Albada, and P Sloot. Towards grid-aware time warp. *Simulation: Transactions of The Society for Modeling and Simulation International*, 81(4):293–306, 2005.

[70] W Jie. *POEMS: A Parallel Object-oriented Environment for Multicomputer Systems*. PhD thesis, Nanyang Technological University, Singapore, 2001.

[71] D.B. Johnson. *Distributed System Fault-tolerance Using Message Logging and Checkpointing*. PhD thesis, Rice University, Texas, USA, 1989.

[72] F. Kuhl, R. Weatherly, and J. Dahmann. *Creating Computer Simulation Systems: An Introduction to HLA*. Prentice Hall, New Jersey, USA, 1999.

[73] T. Lake. Time management over inter-federation bridges. In *Proceedings of the 1998 Fall Simulation Interoperability Workshop*, Orlando, Florida, USA, Sept. 1998.

[74] A.M. Law and W. David Kelton. In *Simulation Modeling and Analysis, 3rd edition*, New York, USA, 2000. McGraw-Hill.

[75] C. Liu and B. Hu. Mutual information based on Renyi's entropy feature selection. *In Proceedings of the IEEE International Conference on Intelligent Computing and Intelligent Systems 2009*, pages 816–820, 2009.

[76] L. Liu, S.J. Turner, W. Cai, G. Li, and B.S. Lee. DDM implementation in hierarchical federations. In *Proceedings of the 2002 Fall Simulation Interoperability Workshop*, Orlando, Florida, USA, Sept. 2002.

[77] L. Liu, S.J. Turner, W. Cai, G. Li, and B.S. Lee. Improving data filtering accuracy in hierarchical federations. In *Proceedings of the 36th Annual Simulation Symposium*, pages 209–215, Orlando, Florida, USA, Sept. 2003.

[78] G. Magee, G. Shanks, and P. Hore. Hierarchical federations. In *Proceedings of the 1999 Spring Simulation Interoperability Workshop*, Orlando, Florida, USA, Mar. 1999.

[79] R. McHaney. Integration of the genetic algorithm and discrete-event computer simulation for decision support. *Simulation*, 72(6):401–411, June, 1999.

[80] C. McLean and F. Riddick. The IMS mission architecture for distributed manufacturing simulation. In *Proceedings of the 2000 Winter Simulation Conference*, pages 1539–1548, Orlando, Florida, USA, Dec. 2000.

[81] Defense Modeling and Simulation Office. In *DoD Modeling and Simulation Master Plan*, U.S. Government Printing Office, Washington, DC, USA, 1995.

[82] Defense Modeling and Simulation Office. In *Software Distribution Center*, 2003.

[83] B. Möller, B. Löfstrand, and Mikael Karlsson. Developing fault tolerant federations using HLA evolved. In *Proceedings of 2005 Spring Simulation Interoperability Workshop*, San Diego, California, USA, Apr. 2005.

[84] A. Morrison, C. Mehring, T. Geisel, AD. Aertsen, and M. Diesmann. Advancing the boundaries of high-connectivity network simulation with distributed computing. *Neural Comput*, 17(8):1776–801, 2005.

[85] K.L. Morse, D. Drake, and R.P.Z. Brunton. Web enabling an RTICAN XMSF profile. In *Proceedings of the 2003 European Simulation Interoperability Workshop*, Stockholm, Sweden, Jun. 2003.

[86] K.L. Morse and M.D. Petty. Data distribution management migration from DoD 1.3 to IEEE 1516. In *Proceedings of the Fifth IEEE International Workshop on Distributed Simulation and Real-Time Applications*, pages 58–65, Cincinnati, Ohio, USA, Aug. 2001.

[87] S.R. Musse and D. Thalmann. Hierarchical model for real time simulation of virtual human crowds. *IEEE Trans. Visualiz. Comput. Graph*, 7(2):152–164, 2005.

[88] M.D. Myjak, D. Clark, and T. Lake. RTI interoperability study group final report. In *Proceedings of 1999 Spring Simulation Interoperability Workshop*, Orlando, Florida, USA, Sept. 1999.

[89] M. Nicola and M. Jarke. Performance modeling of distributed and replicated databases. *IEEE Transactions on Knowledge and Data Engineering*, 12(4):645–672, Jul./Aug. 2000.

[90] A. Okutanoglu and M. Bozyigit. Proximity-aware synchronization within federation communities. In *Proceedings of the Tenth IEEE International Symposium on Distributed Simulation and Real-Time Applications*, pages 185–192, Torremolinos, Malaga, Spain, Oct. 2006.

[91] K Pan. *A Service Oriented HLA RTI on the Grid*. PhD thesis, Nanyang Technological Universtiy, Singapore, 2007.

[92] J. Plevyak. *Optimization of Object-Oriented and Concurrent Programs*. PhD thesis, University of Illinois, Urbana-Champaign, Illinois, USA, 1996.

[93] A. Pope and R. Schaffer. The SIMNET network and protocols. Technical Report 7627, BBN Systems and Technologies, 1991.

[94] M. Roccetti, P. Salomoni, and M.E. Bonfigli. A design for a simulation-based multimedia learning environment. *Simulation*, 76(4):214–221, Apr. 2001.

[95] K. Rycerz, M. Bubak, M. Malawski, and P. Sloot. A framework for HLA-based interactive simulations on the grid. *Simulation*, 81(1):67–76, Jan. 2005.

[96] T. Schulze, S. Straßburger, and U. Klein. Online-data processing in simulation models: New approaches and possibilities through HLA. In *Proceedings of the 1999 Winter Simulation Conference*, pages 1602–1609, Washington DC, USA, 1999.

[97] T. Schulze, S. Straßburger, and U. Klein. HLA-federate reproduction procedures in public transportation federations. In *Proceedings of the 2000 Summer Computer Simulation Conference*, Vancouver, Canada, July. 2000.

[98] G. Schwarz and H.J. Mosler. Investigating escalation processes in peace support operations: An agent-based model about collective aggression. In *Proceedings of the Third Annual Conference of the European Social Simulation Association*, 2005.

[99] O. Shehory, K. Sycara, P. Chalasani, and S. Jha. Agent cloning: An approach to agent mobility and resource allocation. *IEEE Communications*, 36(7):58–67, July, 1998.

[100] W.R. Stevens. *UNIX Network Programming, Networking APIs: Sockets and XTI*, volume 1. Prentice Hall, New Jersey, USA, 2nd edition, 1998.

[101] W.R. Stevens. *UNIX Network Programming, Inter-Process Communications*, volume 2. Prentice Hall, New Jersey, USA, 2nd edition, 1999.

[102] S. Straßburger, T. Schulze, U. Klein, and J.O. Henriksen. Internet-based simulation using off-the-shelf simulation tools and HLA. In *Proceedings of the 1998 Winter Simulation Conference*, pages 1669–1676, Washington DC, USA, Dec. 1998.

[103] K.P. Sycara. Multi-agent systems. *AI Magazine: The American Association for Artificial Intelligence*, pages 79–92, June, 1998.

[104] K.P. Sycara, A. Pannu, M. Willamson, Dajun Zeng, and K. Decker. Distributed intelligent agents. *IEEE Expert*, 11(6):36–46, Dec. 1996.

[105] A.S. Tanenbaum. *Computer Networks*. Prentice Hall, New Jersey, USA, fourth edition, 2003.

[106] A.S. Tanenbaum and M. van Steen. *Distributed Systems: Principles and Paradigms*. Prentice Hall, New Jersey, USA, 2002.

[107] Bernard Testa. Dispersal (entropy) and recognition (information) as foundations of emergence and dissolvence. *Entropy*, 11(4):993–1000, 2009.

[108] S.J. Turner, W. Cai, and B.P. Gan. Adapting a supply-chain simulation for HLA. In *Proceedings of the Fourth IEEE International Workshop on Distributed Simulation and Real-Time Applications*, pages 67–74, San Francisco, California, USA, Aug. 2001.

[109] F. Vahid. Procedure cloning: A transformation for improved system-level functional partitioning. *ACM Transactions on Design Automation of Electronic Systems*, 4(1):70–96, Jan. 1999.

[110] C. Van Ham and T. Pearce. The SIP-RTI: An HLA RTI implementation supporting interoperability. In *Proceedings of the Tenth IEEE International Symposium on Distributed Simulation and Real-Time Applications*, pages 227–234, Torremolinos, Malaga, Spain, Oct. 2006.

[111] Lizhe Wang and Wei Jie. Towards supporting multiple virtual private computing environments on computational grids. *Advances in Engineering Software*, 40(4):239–245, 2009.

[112] Lizhe Wang, Gregor von Laszewski, Marcel Kunze, Jie Tao, and Jai Dayal. Provide virtual distributed environments for grid computing on demand. *Advances in Engineering Software*, 41(2):213–219, 2010.

[113] Y. Xie, Y.M. Teo, W. Cai, and S.J. Turner. Service provisioning for HLA-based distributed simulation on the grid. In *Proceedings of the Nineteenth ACM/IEEE/SCS Workshop on Principles of Advanced and Distributed Simulation (PADS 2005)*, pages 282–291, Monterey, California, USA, Jun. 2005.

[114] M. Xiong, Wentong Cai, Suiping Zhou, Malcolm Yoke Hean Low, Feng Tian, Dan Chen, Daren Wee Sze Ong, and Benjamin D. Hamilton. A case study of multi-resolution modeling for crowd simulation. In *Proceedings of the Agent-Directed Simulation Symposium (ADS'09)*, pages 22–27, San Diego, California, USA, March 2009.

[115] H. Yu and A. Vahdat. Design and evaluation of a Conit-based continuous consistency model for replicated services. *ACM Transactions on Computer Systems*, 20(3):239–282, Aug. 2002.

[116] Z. Yuan, W. Cai, W.Y.H. Low, and S.J. Turner. Federate migration in HLA-based simulation. In *Proceedings of the 2004 International Conference on Computational Science*, pages 225–233, Kraków, Poland, June 2004.

[117] Z. Yuan, W. Cai, and Y.H. Low. A framework for executing parallel simulation using RTI. In *Proceedings of 7th IEEE International Symposium on Distributed Simulation and Real Time Applications (DSRT 2003)*, pages 12–19, Delft, the Netherlands, Oct. 2003.

[118] K. Zając, M. Bubak, M. Malawski, and P. Sloot. Towards a grid management system for HLA-based interactive simulations. In *Proceedings of the 7th IEEE International Symposium on Distributed Simulation and Real Time Applications*, pages 4–11, Delft, the Netherlands, Oct. 2003.

[119] S. Zhou. A trace-driven simulation study of dynamic load balancing. *IEEE Transactions on Software Engineering*, 14(9):1327–1341, Sept. 1988.

[120] S. Zhou, D. Chen, W. Cai, L. Luo, M. Low, F. Tian, V. Tay, D. Ong, and B. D. Hamilton. Crowd modeling and simulation technologies. *ACM Transactions on Modeling and Computer Simulation*, 20(4):Article 20, Oct. 2010.

[121] S.P. Zhou, D. Chen, W.T. Cai, L.B. Luo, Y.H. Low, F. Tian, S.H. Tay, W.S. Ong, and B.D. Hamilton. Crowd modeling and simulation technologies. *ACM Transactions on Modeling and Computer Simulation*, 20(4), 2010.

[122] W. Zong, Y. Wang, W. Cai, and S.J. Turner. Service provisioning for HLA-based distributed simulation on the grid. In *Proceedings of the 8th IEEE International Symposium on Distributed Simulation and Real Time Applications (DSRT 2004)*, pages 116–124, Budapest, Hungary, Oct. 2004.

Index